Amateur Radio INSIGHTS

Understanding Amateur Radio Mathematics

by

Michael Burnette, AF7KB

This book has been published by Radio Society of Great Britain of 3 Abbey Court, Priory Business Park, Bedford MK44 3WH, United Kingdom
www.rsgb.org.uk

First Edition 2024

Amateur Radio Insights is an imprint of the Radio Society of Great Britain

© of this edition is held by the Radio Society of Great Britain & Michael Burnette, AF7KB, 2024. All rights reserved. No part of this publication may be reproduced, stored in a retrieval system, or transmitted, in any form or by any means, electronic, mechanical, photocopying, recording or otherwise, without the prior written permission of the copyright holders.

This book is a combined title based on original works by Michael Burnette, AF7KB called *The Fast Track to Mastering Technician Class Ham Radio Math*, *The Fast Track to Mastering General Class Ham Radio Math* and *The Fast Track to Mastering Extra Class Ham Radio Math* which are part of the author's series of Fast Track books. Copyright for this book and others in that series belongs to Michael Burnette, AF7KB and all moral rights to these works are asserted.

Please Note:
The opinions expressed in this book are those of the author and are not necessarily those of the Radio Society of Great Britain. Whilst the information presented is believed to be correct, the author, the publishers and their agents cannot accept responsibility for consequences arising from any inaccuracies or omissions.

ISBN: 9781 9139 9568 3

Cover design: Kevin Williams, M6CYB
Typography and design: Michael Burnette, AF7KB
Production: Mark Allgar, M1MPA

Printed in Great Britain by CPI Anthony Rowe Chippenham, Wiltshire

Any amendments or updates to this book can be found at:
www.rsgb.org/booksextra

Contents

Contents iii
    Image Credits v
    List of Figures v
Introduction vii
    Tools of the Ham Radio Maths Trade vii
    How to Win the Maths Game viii
    Learning the Theory Behind the Maths x
    Why Do We Learn This Stuff? x
    Maths Is a Powerful Language xi
1 A Quick Review of Some Mathematics Fundamentals 1
    Operations 1
    Types of Numbers 3
    Superscripts & Subscripts 3
    Graphing 4
2 The Smallest Possible Amount of Algebra 9
    The Real Algebra 11
3 The Metric System for Electronics 13
    How To Convert Values 15
    Number Sense 17
    Application 18
4 Wavelengths and Frequencies 21
    Application 26
5 Decibels – The Comparison Value 29
    Application 32
6 Ohm's Law & Joule's Law 35
    Application 37
7 Series & Parallel Calculations 41
    Application 44
    A Shortcut! 46
8 Peak-to-Peak and RMS Voltage 47
    Application 48
9 Receiver Performance Characteristics 51
    Application 53
10 Interference Calculations 57
    Noise Floor 59
11 Resonant Frequency of a Circuit 61
    Application 63
12 Calculating Reactance 67
    Application 71
13 The Q Factor 73
    Application 74

14	**Time Constants**	**81**
	Application	83
15	**Phase Angles**	**87**
	Power Factor	90
	Application	93
16	**Calculating & Graphing Impedance**	**95**
	Impedance Basics	95
17	**Rectangular & Polar Coordinates**	**99**
	Impedance Matching	102
	Application	102
	Impedance Matching With a Transformer	108
	Application	109
18	**Op-Amp Circuits**	**111**
	Application	112
19	**Modulation Index & Deviation Ratio**	**113**
	Application	115
20	**Bandwidth Calculations**	**117**
	CW Bandwidth	118
	Amplitude Modulation Bandwidth	120
	AM Bandwidth Calculations	123
	FM Bandwidth Calculations	123
	Digital Signal Bandwidths	124
	Claude Shannon, Harry Nyquist, & Ralph Hartley	124
	Calculating Digital Signal Bandwidths	126
	Application	126
21	**ERP, EIRP, Link Budget & Link Margin**	**129**
	Application	130
	Link Margin & Link Budget	131
	Application	132
22	**Transmission Lines**	**135**
	Anatomy & Physiology of a Coaxial Cable	137
	Applications	141
	Transmission Lines for Impedance Matching	141
	Velocity Factor	142
23	**The Smith Chart**	**145**
	Application	150
	Coaxial Cable Stubs	150
	SWR Circle	152
	Introduction to VNA's	156
24	**Maxwell's Equations**	**159**
	Gauss's Law for Electricity	161
	Gauss's Law for Magnetism	165
	Faraday's Law of Induction	167
	Ampére's Law with Maxwell's Addition	168
	Electromagnetic Waves	169
Index		**173**
	About the Authors	178

Image Credits

All images either author's own work or Public Domain except:

Full Smith Chart: Cannabic, CC BY-SA 4.0
https://creativecommons.org/licenses/by-sa/4.0,viaWikimediaCommons
https://upload.wikimedia.org/wikipedia/commons/2/2a/Smith_chart.svg

"Cable Making circa 1858": Robert Charles Dudley, CC0, via Wikimedia Commons

List of Figures

1	TI-30XS Calculator	xii
1.1	Number Line	4
1.2	Graph of x = 1	5
1.3	Rectangular Coordinates Graph	5
1.4	Graph of x + y = 1	6
1.5	Creation of A Sine Wave	7
1.6	Sine Function	7
2.1	Muhammad ibn Musa al-Khwarizmi, c 780 – c 850	9
3.1	Metric Values	14
3.2	Setting TI-30XS for Engineering Notation	16
3.3	Conversion of 3,720 mA to 3.72 A	16
3.4	Metric Value Converstion Number Line	17
4.1	One Wavelength = One Cycle	22
4.2	Frequency & Wavelength Relationship	24
5.1	Calculator Keystrokes and Screen for Equation 5.1	30
5.2	Keystrokes for Equation 5.7	33
6.1	Voltage Divider	39
7.1	Series-Parallel Circuit	41
7.2	Analyzing a Series-Parallel Circuit	42
7.3	Ridiculous Resistance Problem	42
7.4	Keystrokes for Equation 7.4	45
8.1	Peak-To-Peak, Peak, and RMS Voltages	48
8.2	Keystrokes for Equation 8.7	49
9.1	Receiver Specifications for ICOM IC-7300 ©2024 Icom America Inc	56
11.1	Values of y = x, y = $\sqrt{x}$, and y = x2	63
11.2	Reactance Values	63
11.3	Resonance Calculation	64
12.1	Inductive and Capacitive Reactances vs. Frequency	67
12.2	Keystrokes for Equation 12.6	69
12.3	Keystrokes for Equation 12.9	70
13.1	Effects of Q on Frequency Response	73
13.3	Parallel Resonant Circuit Frequency Response & Half- Power Bandwidth	76
13.4	Input Signal (L) vs. Signal Produced by Ringing (R)	78
14.1	Time Delay Circuit	82
14.2	Keystrokes for Equation 14.4	84
15.1	Sine Waves 90° Out of Phase	88
15.2	Oliver "Happy" Heaviside	88
15.3	ELI the ICE Man	90
15.4	Two Sine Waves 90° Out of Phase	91
15.5	Real Power at 90° Phase Difference	91
15.6	Capacitor Bank at Electrical Substation	93
16.1	Phasor Diagrams	98
17.1	Polar Coordinates	101
17.2	Rectangular Coordinates	101
17.3	US Amateur Extra Class Exam Figure E5-1	103
17.4	Keystrokes for Equation 17.2	104
17.5	Resistance, Reactance, Impedance, and Phase Angle	107
17.6	Keystrokes for Equation 17.9	108
18.1	Op-Amp	112
20.1	CW Waveform	119
20.2	PARIS Consists of 50 Symbols	120
20.3	Amplitude Modulated Signal	121
20.4	Single Sideband	122
20.5	Keystrokes for Equation 20.7	126
22.1	Window Line	135
22.2	Cable Manufacturing circa 1858	136
22.3	Internal Structure of Coaxial Cable	137
22.4	Equivalent Circuit of Coaxial Cable	138
23.1	Smith Chart	147
23.2	Main Smith Chart Axes	148
23.3	Resistance Circles & Reactance Arcs	149
23.4	Slide Rule	150
23.5	Super-Simple Smith Chart	151
23.7	Plotting the Resistance Part of Z	153
23.6	Phasor Diagram	153
23.8	Steps 2 & 3 Creating SWR Circle	154
23.9	The Impedance is Where The Circle and the Arc Meet	155
23.10	Drawing the SWR Circle	155
23.11	SWR vs. Frequency Plotted by VNA	157
23.12	VNA Smith Chart Plot of Multiple Resonances	158
24.1	James Clerk Maxwell	159
24.2	Charles-Augustin de Coulomb	160
24.3	André-Marie Ampére	160
24.4	Carl Friedrich Gauss	162
24.5	Vector Field of Winds	163
24.6	Vector Field	164
24.7	Vector Magnetic Field	166

Introduction

The language of mathematics is what took the electrical and magnetic effects noticed by early researchers from laboratory oddities to world-changing technology.

You might say that the dance of forces that creates radio speaks fluent mathematics. Much as we admire the brilliance of such figures as Ørsted, Coulomb, Ampére, and Faraday, it was the mathematics of George Simon Ohm and, especially, James Clerk Maxwell that opened up the possibility of turning those curious things that happened in the laboratory into practical technology.

There's a very practical reason to make the effort to familiarize yourself with amateur radio mathematics. In electronics, "plug it in and see if any smoke comes out" is a perfectly valid test method, but one that can get expensive and dangerous; doing a little calculation can save a lot of money and heartache.

Mathematics gives us a rock of certainty from which to reason about electronics. After all, Ohm's Law does not say "V is usually in the neighborhood of I times R." It says, "Voltage, current, and resistance are *always* precisely proportional."

Finally, there's something to be said about developing an appreciation for the elegance and economy of the mathematical language. It takes many words to describe how a radio wave comes into being, propagates, and affects the receiving antenna on the other end. James Clerk Maxwell managed to cover it all in four short equations.[1]

Tools of the Ham Radio Maths Trade

As we work through the applications in this book, we'll show "calculator-key-by-calculator-key" how to perform the calculations for each equation. While any scientific calculator will suffice, we'll be giving specific exam-

---

[1] To be honest, Maxwell covered it in 21 not-very-short equations, but Oliver Heaviside simplified them to the four we use for most purposes today.

ples for the Texas Instruments TI-30XS, like the one shown in Figure 1. There are plenty of calculators out there that use different key strokes than we show, but if you have one of those, we're going to assume you know how to use it.

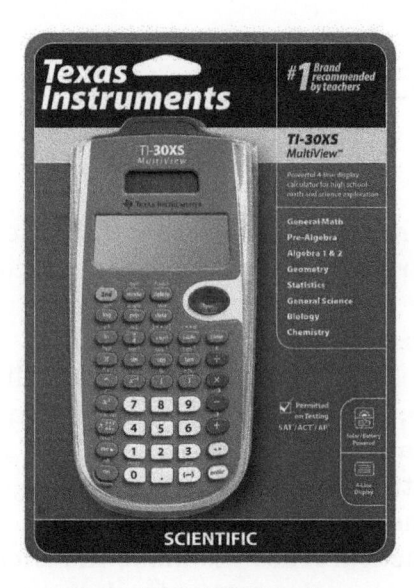

Figure 1: TI-30XS Calculator

It lets you see all the steps of your calculations, is easy to use, has all the functions you'll need (and many more), and is only approximately $20, £25, or €30. You can enter formulas exactly the way they appear in most books, including this one. They're also durable – let's face it, they designed the thing to survive in one of the most unimaginably brutal environments on Earth – a student's backpack! If you purchase a TI-30XS, peek at the instructional video on fasttrackham.com for some tips on using it.

https://fasttrackham.com/2020/05/09/intro-to-the-ti-30xs-calculator/

As you prepare to march forward into this battle, we also suggest you arm yourself with an appealing notebook and a generous supply of the writing instrument of your choice.

How to Win the Maths Game

If you are part of the estimated 20% of people who experience genuine anxiety whenever maths is so much as mentioned, here are a few things to know about what's covered in this book that might help you relax:

- This is simple maths. Until we get to the highest levels, it is all add, subtract, multiply and divide. You plug a couple of numbers into a simple formula, press the right keys on your calculator, and the answer falls out like magic. Even at the highest levels of the calculations we will cover, which barely touch the simplest parts of trigonometry, it is a matter of pushing the correct calculator buttons. No algebra, no quadratic equations, no calculus, no statistics required. The trick is in remembering what formula to use – and that's where **practice** with the applications in this book comes in.

- Do you consider yourself to be "not a maths person?" Today's neuroscience seems to show that, with rare exceptions, no one starts life as "a maths person" or "not a maths person." The only real difference between "maths people" and "not maths people" is the "maths people" *have practiced a lot of maths*. (Aw, rats!) So, practice, practice, practice and it gets easier and easier.

- **There is no substitute for practice.** *Don't just read the formulas, do your best to work through each problem,* with pencil and paper. There's something very powerful about the brain-hand-eye connection that helps us learn faster and more deeply than just punching the buttons on a calculator.

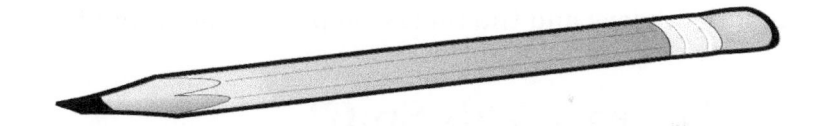

- If something still isn't making any sense, *try again*. Very, very few people "get" this stuff the first time. You must wrestle with it. Master the basics and you will be very glad you did as you progress in the hobby.

- Here's a little secret: If you are using this book to study for an amateur radio exam *anywhere in the world*, chances are very good that if you manage to miss *every single maths problem* on the exam *for any level*, but get everything else right, you'll almost certainly still pass. You won't be much of a ham radio operator, though. So, take a deep breath, relax, and learn. Remember – ham radio is a *hobby*, it's supposed to be *fun!* We learn much better when we're relaxed and enjoying ourselves – and being relaxed *and* having fun are both your choice.

We realize that for those with little or no electronics background, the maths of ham radio can look more than a little daunting. However, if you're willing to put in some work, there is absolutely no reason why the maths should stop you from getting any license you want.

Learning the Theory Behind the Maths

Most people, certainly including us, find it much easier to understand all this mathematics when they understand the vocabulary and theory behind it. This really came home to Michael back in the days just before the *Fast Track* series was born, when a friend of his was struggling mightily to come to grips with Ohm's Law. Ohm's Law is a set of simple equations that are used to calculate the relationships among the fundamental electrical values of volts, amps, and ohms. This struggle seemed unlikely, to say the least. This was a highly intelligent person with a background in maths and science – how could this person be struggling?

Finally, in complete exasperation, his friend demanded, "WHAT'S A VOLT?" Oh! Michael went through a quick explanation of voltage, current, and resistance, and it all clicked.

We won't be going deeply into electronic theory in this workbook, since that's covered in considerable depth in other sources, but we suggest really learning the theory. Believe it or not, the maths helps you understand the theory and the theory helps you understand the maths.

Why Do We Learn This Stuff?

> *[The universe] cannot be read until we have learnt the language and become familiar with the characters in which it is written. It is written in mathematical language, and the letters are triangles, circles and other geometrical figures, without which means it is humanly impossible to comprehend a single word.*
> *– Galileo*

To be bluntly honest with you, if you don't know the maths of ham radio, you don't know ham radio. In a very real sense, electronics *is* maths.

Some national testing systems, such as the UK's, include a quite civilized element that is lacking from the US testing rooms; examinees are provided with a pamphlet containing the relevant formulas for the exam as well as some additional reference material. Of course, you will still need to know which formula to use to solve each question, but at least you'll have a memory aid at hand.

What size fuse do you need to put in this circuit to keep it safe? What transformer do you need to match this transmitter to that transmission line? How long should you make your homemade antenna? How much do you need to raise the power of your transmitter to get that ham in Slovenia to be able to hear you? How good is this transceiver you're thinking about buying? The answers to all those questions and many more are in the maths.

People just entering the hobby often don't realize that their license isn't just a license to pick up a handheld radio, push the button, and transmit a signal. It's a license to legally do a *lot* more, including to build *and operate* your own radio equipment without ever having to have it checked by the official Powers-that-be. That's more rights than the big radio manufacturers have! In return for all that freedom, the authorities asks that we at least know the very basics of electricity. It's not such a bad deal, really.

While ham radio is usually a very safe hobby, there are aspects of it that can be, and are, quite hazardous. Part of controlling the danger is knowing the maths to predict what is going to happen when we switch on the power.

Ham radio can also interfere with other types of radios – imagine being the person who accidentally interferes with the communications gear and electronics on a Boeing 787 passenger jet trying to land. The results could be very bad indeed. We never want to create interference to other services, and, in fact, it's quite illegal to do so. Part of the science of not interfering with other signals is knowing the maths of our hobby.

Aside from the considerable practical benefits of mastering the mathematics of amateur radio, there is what some call the beauty and wonder of mathematics. In the same way we can find beauty in a Bach concerto or a Rembrandt painting, we can find beauty in mathematics. If nothing else, one can marvel at the minds that came up with these things.

Maths Is a Powerful Language

To at least one of your authors (the one who didn't become a maths teacher!) and maybe to you, maths in school seemed like nothing but a bunch of arbitrary procedures. "Add these two numbers together, carry the 1, add those three numbers together ....", "Multiply both sides of the equation by this number to solve for x ....", etc. It can seem to be as senseless as the directions for what we in the US call the Hokey Pokey: "Put your right foot in, put your right foot out, put your right foot in and shake it all about."

All those procedures are how you do *calculations*, but they are not what mathematics *is* – and sometimes, no one mentions that to us on our way through school, so we never quite get *why* we're solving for x. We will do our best to illuminate the "why's" in this work.

Maths is a language, one that packs a lot of meaning into very few symbols. It's almost like poetry!

As beginning hams, we learn Ohm's Law. It's a simple formula:

$$V = I \times R$$

Formulas don't get much simpler than that one. It says "**Voltage**, **current**, and **resistance** are all proportional to each other." At this level, we can think of an electrical circuit as a garden hose. In this model, water pressure is the voltage. The amount of water flowing through the hose is the current, and your thumb over the end of the hose is the resistance.

That formula describes a complicated relationship among those three fundamental values of electronic circuits. It says, "These three values are all related to one another in this way: If we increase the **voltage**, more current will flow. If we increase the **resistance**, less **current** will flow, and all of those changes are proportional to each other." Not only that, it also says that **current** through the circuit is equal to the **voltage** across the circuit divided by the **resistance** in the circuit, *and* it says the **resistance** in a circuit is equal to the **voltage** across the circuit divided by the **current** running through the circuit. In just five symbols, Ohm's Law manages to say what it took us over 500 characters (counting spaces) to say in English.

The mathematical language of Ohm's Law is simple compared to that of Maxwell's Equations. Here they are in what mathematicians call "differential form."[2]

$$\vec{\nabla} \cdot \vec{E} = \rho_f$$
$$\vec{\nabla} \cdot \vec{B} = 0$$
$$\vec{\nabla} \times \vec{E} = -\frac{\partial \vec{B}}{\partial t}$$
$$\vec{\nabla} \cdot \vec{B} = \mu_0(\vec{J} + \frac{\partial \vec{E}}{\partial t})$$

---

[2] More precisely, *partial differential form."* That ∂ is the sign for *partial derivative*. Put as simply as possible, it just means the variable marked with the ∂ in the equation is changing while the others are treated as constants.

Of course, with those equations we have now gone far beyond "+, −, ×, and ÷," but don't concern yourself with any symbols in there that are foreign to you. Just recognize that in a mere 35 symbols, those four equations describe the fundamental truths of almost everything we know about electromagnetic radiation as well as serving as the seeds of Einstein's discoveries in relativity. We'll cover the basics of what those equations say in the last chapter of this book, but it has been over 160 years since Maxwell published them and mankind is still discovering all of the implications of those brilliant equations.

Putting lots of meaning into very little space and doing so very clearly, precisely, and usefully is what maths is really all about. It's also why maths can be a little challenging to understand.

The lesson here, is have patience. It takes some time to really understand these formulas *because there is a lot in them*, not because they're inherently "hard." Albert Einstein himself did not fully grasp all the implications of his own famous formula, $E = mc^2$, so it's not reasonable to expect yourself to instantly grasp all the implications of Ohm's Law, much less some of the more complex equations. That would be like expecting yourself to read all of *War and Peace* with a glance at the cover.

The good news is that, unlike some languages, maths is designed to be as easy to understand as possible, given how much information it is carrying. Compare the clarity of that Ohm's Law formula with, say, the English sentence, "I saw a man on a hill with a telescope."

That seems simple enough, but what does it mean, precisely?

- There's a man on a hill, and I was seeing him through my telescope?

- There's a man on a hill, whom I was seeing, and *he* has a telescope?

- I saw a man, and he was on a hill that also had a telescope on it?

- I was on a hill, and I was seeing a man who was using a telescope?

- There is a man on a hill, and I'm now sawing him in half with a telescope?

- .....and there are more ...

Mathematics, though, is a precise language. Once you know the language of maths, there is only one possible meaning for the Ohm's Law formula or any other formula. That's one of the main superpowers of the language called maths.

Speaking of languages; we are both Americans. We are well aware of the old joke about "two nations divided by a common language." Even the very subject of this work has one name on one side of the Atlantic and another on the far shore. Americans shorten *mathematics* to *math* while our cousins across the briny waves politely point out that the subject is not, after all, *mathematic* so they shorten it to *maths*. (We must admit they have a point that is difficult to refute.) We have attempted to translate our Americanese into proper English for our international readers. It seems unlikely that we have completely succeeded, so we beg your indulgence for any linguistic missteps.

Let's go challenge the math(s) dragon!

Chapter 1

A Quick Review of Some Mathematics Fundamentals

Operations

Operators tell us what to do with numbers. If numbers (and symbols such as x and π) are the nouns of the language of maths, then operators are the verbs. Some can be written several different ways. Relation symbols, such as the equal sign (=) are the adjectives.

Using symbols like x or even V for *volts* allows us to generalize mathematical principles. It is all well and good to know that $3^2 + 4^2 = 5^2$, but it is much more powerful to know that the sides of a right triangle, called *side a, side b and side c* with c being the long side (the "hypotenuse") *always* have the relationship $a^2 + b^2 = c^2$.

There are *many* more operators in the language of maths, but most of them are just elaborations on addition, subtraction, multiplication, and division. Exponents and roots are just forms of multiplication and division, and multiplication and division are just fancy addition and subtraction!

Even in calculus – which is *way* beyond any of the maths in any of the license levels for any country we know of – there's $\sum$, the "sum" operator which basically means, "do a bunch of maths and then *add* up the results." A considerable amount of the content of calculus courses involves the various shortcuts to accomplish tasks like that quickly, but in the end, it's still addition.

\multicolumn{4}{c	}{**Common Operators & Symbols**}
Symbol	**Form**	**Meaning**	**Example**
$=$	$x=4$	Equals. "Is the same as."	$2+4=6$
$+$	$x+y=z$	Add the value represented by x to the value represented by y.	$3+2=5$
$-$	$x-y=z$	Subtract the value represented by y from the value represented by x	$3-2=1$
$\times$	$x \times y = z$	Multiply x by y. In other words, add x to itself y times.	$3 \times 2 = 6$
$\cdot$	$x \cdot y = z$	Multiply x by y.	$3 \cdot 2 = 6$
$()()$	$(x)(y)$	Multiply x by y.	$(3)(2)=6$
No symbol	$xy=z$	Multiply x by y. Only used if at least one of the items is a symbol, not a number.	$3x=6$
$\div$	$x \div y = z$	Divide x by y. In other words, "you can subtract y from x z times."	$4 \div 2 = 2$
$/$	$x/y=z$	Divide x by y.	$4/2=2$
$\frac{numerator}{denominator}$	$\frac{x}{y}=z$	Divide x by y.	$\frac{4}{2}=2$
$value^{value}$	x^y	Exponent. Multiply x by itself y times.	$4^2=16$
$\sqrt{value}$	$\sqrt{x}=y$	Square root. What number multiplied by itself equals x?	$\sqrt{4}=2$
$\approx$	$x \approx y$	"x is approximately equal to y."	$\frac{3}{5} \approx 1.67$
$<$	$x<y$	x is less than y.	$2<4$
$<$	$x>y$	x is greater than y.	$4>2$
$\leq$	$x \leq y$	x is less than or equal to y.	$x \leq 4$
$\geq$	$x \geq y$	x is greater than or equal to y	$x \geq 4$
Ω	$R_1 = 100\Omega$	Symbol for Ohms	$volts = \frac{amps}{\Omega}$
Δ	Δ_f	The change in value of, in this case, f.	$\Delta_f = 3\,kHz$
λ	$\lambda = 2\,m$	Common symbol for wavelength	$\frac{300}{\lambda}$

Table 1.1: Mathematical Operators

Types of Numbers

Numbers can be of several different types.

- Positive Numbers: Numbers greater than zero. $\frac{1}{2}, 1, 2, \sqrt{4}.etc.$
- Negative Numbers: Numbers less than zero. $-\frac{5}{8}, -\pi, -42$
- Natural Numbers: Positive integers that start from 1. 1, 2, 3, 4, etc.
- Whole Numbers: Positive integers that start from zero. All the natural numbers plus the number zero.
- Integers: Zero, positive and negative numbers without fractions or decimals. $-3, -2, 0, 2, 4$, etc.
- Rational Numbers: Numbers that can be expressed precisely as a fraction, $\frac{x}{y}$. $-4, 1, 2\frac{1}{2}, 0.58$.
- Irrational Numbers: Numbers that cannot be expressed precisely as a fraction. $\pi, \sqrt{2}$.
- Real Numbers: All the rational numbers and irrational numbers. All these numbers can be located on a number line. $-\frac{3}{4}, \frac{4}{5}, -3, \pi, 42$.
- Imaginary Numbers: Numbers that cannot be located on a number line. $\sqrt{-1}$
- Complex Numbers: Numbers that can only be expressed as a combination of a real number and an imaginary number. $x \times \sqrt{-1}, Z = 42, -j310$

A number can occupy more than one of those categories. For instance, π is a positive number and an irrational number.

Superscripts & Subscripts

Superscripts are usually numbers that appear slightly above and to the right of a value in a formula. For instance, in 7^2, the 2 is the superscript; it means to square the 7. To square the 7 we multiply that 7 by itself to get 49. A superscript might also be a symbol such as x, π, or θ. There may be some weird exceptions, but generally speaking a superscript demands some sort of numerical operation. The non-numeric symbols either represent some constant (such as π), a given value, or the result of some other calculation.

Subscripts are for convenience and clarity. They are characters or even words that appear slightly lower and to the right of a value, such as X_C, where the X is, in this case, a value of Capacitive reactance. When you are working on paper you can use any subscripts you like to keep track of values. Frankly, they are also often used to make formulas more compact for typesetting purposes!

One letter you might see in a subscript, a superscript, or even as a part of an equation is n. n just means, "some relevant number."

Graphing

One important maths tool that is used a lot in the license courses and in electronics in general is graphing. It is very useful to know at least the basics of how graphs work.

By giving us a picture of what is going on with a process, graphs help us much more easily understand the underlying forces as well as the results we can expect.

There are many types of graphs, but for the moment the only type that will concern us is known by various names; an "xy graph", a "Cartesian coordinates graph", or a "rectangular coordinates" graph. All those names for that type of graph mean the same thing.

A graph is a picture of an equation. An equation is a mathematical expression that includes an equal (=) symbol. A very simple equation would be $x = 1$. To graph that, we'd start with a number line, which is a way of representing numbers by placing them on a horizontal line, as shown in Figure 1.1.

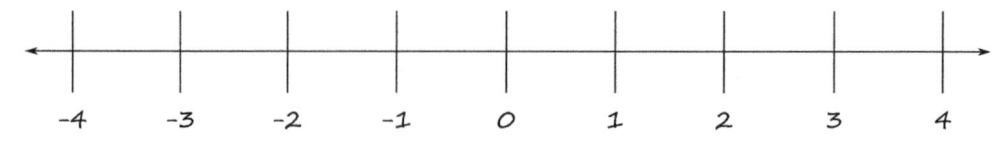

Figure 1.1: Number Line

In graphing, that number line is known as the "x axis" and it shows values for x. (Of course, we could use any letter we wanted, but x is traditional.)

It's pretty simple to graph $x = 1$. We locate 1 on the x axis and make a dot, as we've done in Figure 1.2. We're done!

If you listen carefully, you might notice you hear absolutely no applause for this feat. Plotting a single number on a number line isn't very useful.

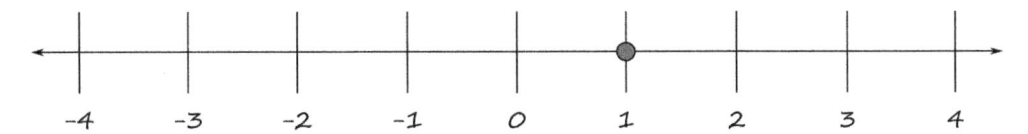

Figure 1.2: Graph of $x = 1$

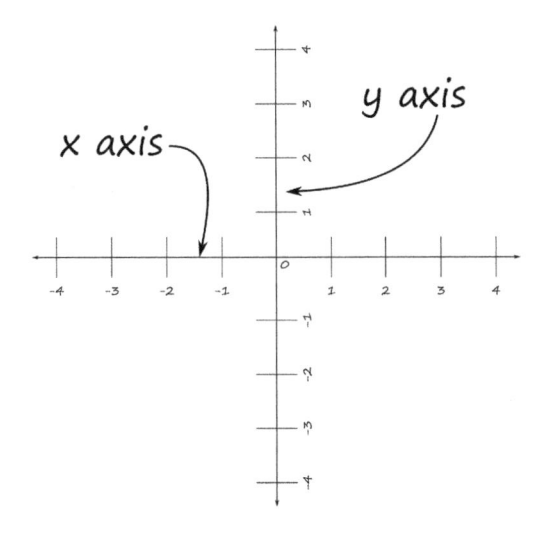

Figure 1.3: Rectangular Coordinates Graph

Once we add another axis to the graph, which is called the y axis, as in Figure 1.3, the graph gets much more useful.

Now we can show the relationship between two or more values. For instance, let's consider the simple equation $x + y = 1$.

There's no single solution for that equation – in fact, it has an infinite number of solutions. We could create a table showing some possible solutions, like Table 1.2, but we'll never get that table filled out completely. Not only does that table fail to tell us anything about what y might be if $x = 1,000,001$, it can't even tell us for sure what y would be if $x = 0.000000000001$.

However, there's a way to show the set of possible solutions, and it looks like the line across the graph in Figure 1.4.

The x axis on that graph is showing us all the possible values of x, or at least, as many as will fit on the page. The y axis is showing us all the possible values of y. To read that graph, you can start on the x axis. Let's say we start at 0. To find the value for y when $x = 0$, we travel up the y axis until we get to that diagonal gray line. Where the gray line crosses the y axis, that's the value for y. The graph shows that for $x + y = 1$, if $x = 0$, $y = 1$.

Possible Solutions for x + y = 1		
x	**y**	**Total**
0	1	1
-1	2	1
-2	3	1
-1,000,000	1,000,001	1
1,000,000	-999,999	1

Table 1.2: Five Possible Solutions for $x + y = 1$

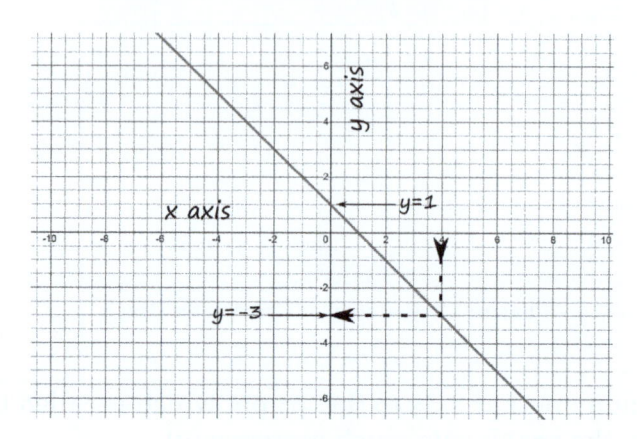

Figure 1.4: Graph of $x + y = 1$

What if $x = 4$? We go straight *down*, toward the gray line, into negative y values, starting at the 4 on the x axis. Where the gray line crosses the line coming straight down from the four, we look to the left at the y axis and see that if $x = 4$, then $y = -3$.

One graph you'll see over and over as you study electronics is a graph of what's called a sine wave. Of course, we'd be perfectly correct if we simply said that the output of some device takes the form of $y = sin(x)$ but a graph is far more informative.

You could think of a sine wave as a map of something's positions at various times as it travels around a circle. On this graph, the x axis represents time, while the y axis represents the amplitude – put simply, the strength – of something that is varying in a cyclic way. On the graph in Figure 1.5, we've made the y axis represent the height above a horizontal line passing through the center of the circle. Values above that line are shown as positive while values below it are negative.

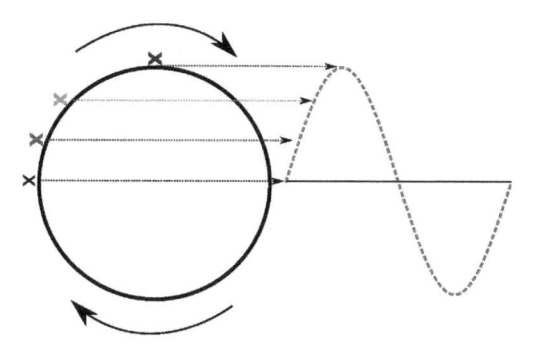

Figure 1.5: Creation of A Sine Wave

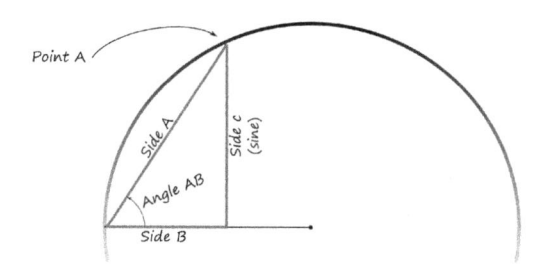

Figure 1.6: Sine Function

You can see that a 9 o'clock, which is the time at which we're starting, we were at "zero height from the horizontal center line of the circle." We plot a mark at "9:00, 0 height." As we travel around the circle, we keep plotting in the same way. At 12:00, we're at the top of the wave and start heading back down. At 3:00, we get back to zero and head into negative territory. When we get back around to 9:00, we're back at zero again. Connect all the dots and it creates the graph we call a sine wave.

Why Is It Called a "Sine" Wave?

The equation that produces that graph of a sine wave is $y = \sin(x)$, sin being short for sine; but that doesn't really answer the question. We're not going to go too deeply into this, but take a look at Figure 1.6.

We've built a triangle in the circle with one point at our 9 o'clock mark and the other at a random point we'll call Point A. We've done that to measure the distance from the horizontal center line of the circle to point A using trigonometry. With that brand of mathematical wizardry, if we know the length of two sides of that triangle *and* we know the angle marked Angle AB, we can figure out the length of Side C.[1] (Just assume we know those lengths and that angle.) The trigonometric function used

---

[1] The trigonometry magic also works if we know two angles and the length of one side.

to do that is called the sine. If we build about a million of those triangles measuring lots and lots of different Point A's, and graph the results with the x axis representing degrees around the circle and the y axis representing the length of Side C, we end up with that graph that we call a sine wave.

If you find yourself overcome with an urge to play with graphing various equations, there's a free graphing calculator at:

https://www.desmos.com/calculator

Chapter 2

The Smallest Possible Amount of Algebra

There is no "real" algebra on any license exam we have ever seen. Okay, with a bit of a stretch we *could* call some of it algebra, just because it uses some letters instead of numbers in the formulas, but if it is algebra, it is just barely algebra.

However, some of what you will learn is *based on* algebra, so we present this brief lesson just so you can understand how we "got from here to there" when the time comes, if that is helpful for you.

Algebra was probably invented by an Iraqi mathematician named Muhammad ibn Musa al-Khwarizmi. Algebra's original name, *al-jabr*, means "the restoration" or "the reintegration" in Arabic. Whether he invented it or not, Al-Khwarizimi wrote a book on the subject sometime near 830 AD. His work was translated into Latin and introduced to Europe around 1500 AD. His name lives on in the English word *algorithm*.

Figure 2.1: Muhammad ibn Musa al-Khwarizmi, c 780 – c 850

First, consider this equation.

$$2 \times 4 = 8 \qquad (2.1)$$

If that is true (and it is ...) then it is also true that:

$$\frac{8}{2} = 4 \qquad (2.2)$$

and...

$$\frac{8}{4} = 2 \qquad (2.3)$$

These three numbers are in a relationship that we might diagram by writing, basically, all three equations at once, like this:

$$\frac{8}{4 \mid 2} \qquad (2.4)$$

That's a *proportional* relationship. Think of the 4 and the 2 working together to produce the 8, like a little machine that produces 8's. The 8 and the 4 are a little machine that makes 2's, etc.

If we know *any* two of those numbers, we can quickly learn the formula that will give us the value of the third by covering up the unknown number. If we cover up the 8, we're left with 2×4. If we cover up the 4, we're left with $8 \div 2$.

So far, so what. But here's where the Mathematical Millenial Falcon goes into hyperdrive.

We can generalize what we have just done into something like this: $x = y \times z$. Then we can make a memory aid that looks like this:

$$\frac{x}{y \mid z} \qquad (2.5)$$

If we have two numbers that we know are proportional to a third number, we can find that third number the same way we did with 8, 4, and 2. Now we have a little machine made of y and z that produces x's. Or it can be a machine made of, say, x's and y's that produces z's.

Really understanding this principle will greatly simplify your journey through Ohm's Law.

The Real Algebra

This little bit of algebra is completely optional. It is only here for the curious, who wonder if there is some formal method of transforming those equations into the various forms we saw. There is, and it goes like this.

We start with $x = y \times z$.

Algebra says we can pretty much do anything we want, mathematically speaking, to any equation, so long as we do it equally to both sides. The "sides" are defined by the $=$ sign. This allows us to simplify equations and even transform them. In this case, we'll divide both sides by y.

$$\frac{x}{y} = \frac{y \times z}{y} \qquad (2.6)$$

Since $\frac{y}{y} = 1$, no matter what the value of y might be[1], we can just ignore that part of the right-hand side. That leaves us with:

$$\frac{x}{y} = z \qquad (2.7)$$

Ta da! $x = y \times z$ transformed into $z = \frac{x}{y}$ in two easy steps. Of course, a similar operation can transform $x = y \times z$ into $y = \frac{x}{z}$ just as handily.

$$\{x = y \times z\} = \{\frac{x}{z} = \frac{y \times z}{z}\} = \{\frac{x}{z} = y\} \qquad (2.8)$$

Finally, please note that the sides of the equation can be reversed. In other words, everything on the left side of the equal sign can move to the right side and vice versa, so:

$y \times z = x$ is the same as $x = y \times z$

$\frac{x}{y} = z$ is the same as $z = \frac{x}{y}$

$\frac{x}{z} = y$ is the same as $y = \frac{x}{z}$

It's the same idea as the way we can say, "Kerry is KC7YL" or "KC7YL

---

[1]Unless $y = 0$. Officially, $\frac{0}{0}$ is "undefined." That's math's polite way of saying it is meaningless rubbish! It's why if you try to divide by zero on any calculator or computer, you'll get an error message.

is Kerry." Both statements are true or, to be logically accurate, both statements are false, but "only one of those is true" is not possible.

Chapter 3

The Metric System for Electronics

Metric prefixes let us express very large and very small numbers easily, and there are a lot of very large and very small numbers in electronics. Values that do not have a metric prefix are known as standard values, basic values, or, sometimes, SI (for Système International) values.

Table 3.1 shows the metric prefixes and their meanings. Figure 3.1 is an attempt to give you some sense of the scale of these values, using a "square" as a standard value. That "square" could be a volt, an ohm, or any other electrical value; the metric prefixes would still mean the same things. Unfortunately, an accurate portrayal of the scales involved would require a square approximately 15,000 miles wide to portray a "gigasquare" and a square approximately one trillionth of an inch wide to show a picosquare. That would be about the size of a single electron.

\multicolumn{5}{c	}{Metric Prefixes}
Prefix	Symbol	Example	Multiplier	Engineering Notation
Giga	G	GHz	1,000,000,000	10^9
Mega	M	MHz	1,000,000	10^6
kilo	k	kV	1,000	10^3
\multicolumn{3}{c	}{Standard Values}	1	10^0
deci	d	dB	0.1	10^{-1}
milli	m	mA	0.001	10^{-3}
micro	μ	μH	0.000001	10^{-6}
nano	n	nF	0.000000001	10^{-9}
pico	p	pF	0.000000000001	10^{-12}

Table 3.1: Metric Prefixes

Metric values always mean the same thing, whether they are applied to electrical units, volume, weight, distance, temperature, or even brightness. A millivolt is a thousandth of a volt, a millimeter is a thousandth of a meter, and a milligram is a thousandth of a gram. If there was such a thing as a millisquare, it would be a thousandth of a square. 500 milligrams is a half a gram and 500 milliamps is a half an amp.

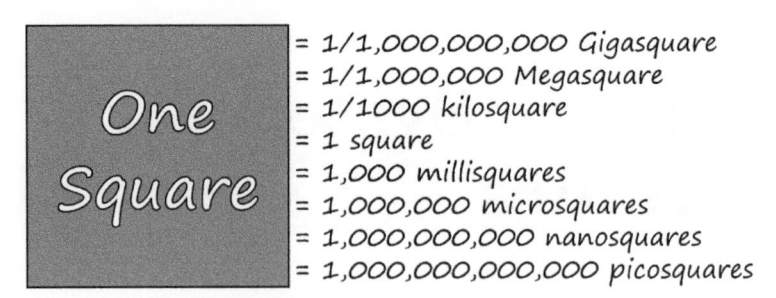

One Square
= 1/1,000,000,000 Gigasquare
= 1/1,000,000 Megasquare
= 1/1000 kilosquare
= 1 square
= 1,000 millisquares
= 1,000,000 microsquares
= 1,000,000,000 nanosquares
= 1,000,000,000,000 picosquares

Figure 3.1: Metric Values

You'll notice the right hand column of Table 3.1 lists the *engineering notation* equivalents of the multipliers for each prefix. Engineering notation is just a variation on scientific notation, also known as *powers of ten*. Both use *exponents*, which are those little numbers tagged onto the upper right hand corner of other numbers. In 10^9, the ^{9} is the exponent. The 10 is called the *base*.

Every number has an exponent. We just don't show them most of the time. In the case of the numbers we usually see in day to day life, the exponent is 1. $10^1 = 10$, $-7^1 = -7$, and $1,747,395^1 = 1,747,395$. There's no point in writing those exponents, but they are still there.

An exponent says, "take that base number and put it into a multiplication problem as many times as the exponent says." In other words,

$$10^2 = 10 \times 10 = 100.$$
$$10^9 = 10 \times 10 \times 10 \times 10 \times 10 \times 10 \times 10 \times 10 \times 10 = 1,000,000,000$$

10^{-2} does *not* equal negative 100. It equals $\frac{1}{10^2}$ or $\frac{1}{100}$, or 0.01.

To express a number that is not an even power of 10, such as 10, 100, 1000, etc., we can use a multiplier – known formally as a *coefficient* – and the appropriate power of 10. 37,200 can be written as 3.72×10^4. The rule for scientific notation is that the multiplier should be equal to or greater than 1 and less than 10, so 3.72×10^4 is correct for scientific notation.

Engineering notation is different from scientific notation only in that it requires all exponents to be divisible by 3; $10^3, 10^6, 10^9$, etc. Of course, this means we throw out the scientific notation rule about the

14

multiplier being "equal to or greater than one and less than ten." 37,200 in engineering notation would be 37.2×10^3. The benefit of engineering notation is that it matches up perfectly with those metric prefixes which makes our conversions really easy. The sole exception is the oddball of the bunch, which is *deci*, but in the ham radio world, that is only used for the value *deci*bels, and we very, very seldom convert those to any other value. (Even a nuclear explosion only creates a momentary sound pressure level of around 250 dBa. Add just a few more dB to that and you're talking about energy levels in the *Star Wars* Death Star range.)

The same rules apply to negative exponents. In scientific notation, we'd write .0372 as 3.72×10^{-2}, but in engineering notation it would be 37.2×10^{-3}.

You might notice that positive exponents tell us how many zeroes there are to the right of the 1. So, $10^3 = 1,000$, and 10^{64} would be a 1 followed by 64 zeroes.

Negative exponents for powers of 10 tell us how many digits there are to the *right* of the decimal point, with the last digit being a 1. So, $10^{-3} = 0.001$.

What about 10^0? It follows the same rules about digits to the left and right of the decimal point – but there's no decimal point! That means $10^0 = 1$. In fact, *any* number raised to the power of 0 equals 1.

Besides being standard practice in the science and engineering worlds, engineering notation and the rules about digits and decimal points will make doing these conversions very simple.

How To Convert Values

We'll call those values with prefixes like "milli" and "kilo" *metric values*.

To convert metric values to standard values, *multiply* the metric value by the power of 10 value for that metric value. To convert from milliamperes to amperes, multiply the number of milliamperes by the power of 10 for milli, 10^{-3}. Multiplying by 10^{-3} means you are multiplying by a fraction ($\frac{1}{1000}$), so you're going to end up with fewer amperes than milliamperes – and that's exactly what you want. (Multiplying by 10^{-3} is the same as dividing by 10^3.)

To convert standard values to metric values, *divide* the standard value by the scientific notation power of 10 value of the metric value. To convert amperes to milliamperes, divide the number of amperes by 10^{-3}. Dividing by 10^{-3} will produce an answer that shows a lot more milliamperes than amperes and that, again, is exactly correct.

You're welcome to haul out your calculator to do those divisions and multiplications, but there's a much, much simpler way to handle these.

Let's say we need to convert 3,720 milliamps to amperes. We could use our calculator to figure out this multiplication problem:

$$3,720.0 \ mA \times 10^{-3} = 3.72 \ A \tag{3.1}$$

To use the TI-30XS calculator for this, you'd use the keystrokes shown in Figure 3.2. We'll use the $\times 10^n$ key to save some strokes.

Before we start, let's set the TI-30XS to give us our answers in engineering notation. That usually makes things simpler.

Figure 3.2: Setting TI-30XS for Engineering Notation

At the end of those keystrokes you'll have a blank screen.

Note that you will press the right arrow key twice in a row.

Now we can proceed with the conversion as shown in Figure ?? .

Notice the first time you pressed Enter the answer came back as $\frac{93}{25}$. In Mathprint mode, The TI-30XS will give you answers like that because those are what you would use to solve, say, an algebra problem. It is a matter of pressing that ⟨▸⟩ key, the *answer toggle* key, to swap the answer into decimal form. You can switch your calculator to Normal mode by following similar steps to how you switched it to engineering notation and it will give you decimal answers instead of the occasional fraction.

We'll show you the precise keys to press on the TI-30XS when we're introducing a new way to use the calculator or for the more complicated equations, but we're going to trust that you can master the simpler calculations on your own – whether with the calculator or by hand.

Notice, though, what happened to the decimal point in 3,720.0. It moved to the left three places – which matches the exponent of 10^{-3}.

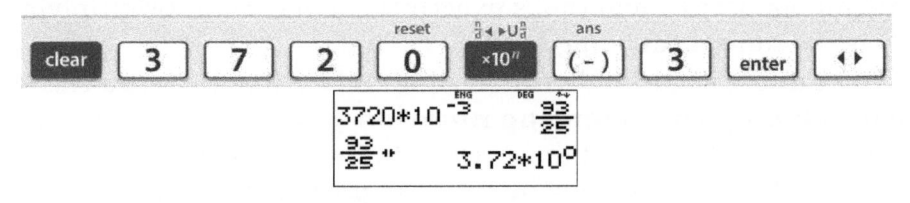

Figure 3.3: Conversion of 3,720 mA to 3.72 A

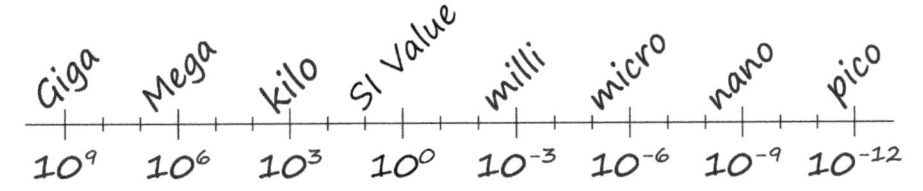

```
Giga   Mega   kilo   SI Value   milli   micro   nano   pico
 |      |      |        |        |       |      |      |
10⁹    10⁶    10³     10⁰      10⁻³    10⁻⁶   10⁻⁹   10⁻¹²
```

Figure 3.4: Metric Value Converstion Number Line

That will always work, no matter what the power of 10 is.

Converting any metric value to another metric value is as simple as moving the decimal point the correct number of places. You don't even need to remember if you're dividing or multiplying, so long as you're clear about the exponent and whether you want to end up with a bigger or smaller number. The number line in Figure 3.4 can help you visualize the values.

On that number line, the left side of the line has the big values. The tiny values are on the right. (There's a method to this madness.)

Ah, but how to remember those metric values? It's not as daunting as you might think. First, remember that except for "deci", which we're ignoring, all the exponents can be divided by three. -12, -9, -6, -3, 3, 6, and 9 all divide evenly by three.

To keep track of those prefix values, memorize this shocking phrase about insane jealousy in the ham community: "**PAUL'S 12 NEW MICRO**PHONES **MADE DESI KILL O**LD **M**AD **GEORGE**." The "P" of Paul stands for pico. The 12 reminds you that pico is 10^{-12}. "New" stands for nano, "*micro*phone" for micro, "made" for milli, "Desi" for deci, "kill old" for kilo, "Mad" for mega, and "George" for giga. The exponent values start at -12 and move up by threes through -9, -6, and -3 until we get to the odd one, deci at -1, then 3, 6, and 9 for kilo, mega, and giga.

> Except for anything the examiners might give you, most examinations do not allow any paper on the table while you take the exam. However, there's no rule that says you can't *make* your own chart of values on the back of your answer sheet while you take the exam.

Number Sense

We'll get to the mechanics of making these conversions, but it is helpful to have some of what the maths teachers call *number sense* about these values.

If you convert 10 volts to kilovolts and come up with 10,000 kilovolts, that should leap off the page and scream something like THAT CAN'T POSSIBLY BE RIGHT!

You have probably seen the game where the object is to guess the number of jellybeans in a jar.

That's a lot of jellybeans. Let's say that jar equals a giga-something. It contains a LOT of all the smaller values, like the milli-somethings, and so on.

If the jar was filled with baseballs, it would be a lower total. The baseballs would be like the standard values or the "big" values like kilo, mega, and giga.

Values that start with pico, nano, micro, and milli are like jellybeans – it takes a lot of them to fill up one standard value. If you are converting from a standard value to one of the smaller-than-standard values, you should end up with a bigger number. On the other hand, if you are converting from one of those "jellybean" values to a standard value or kilo, mega, or giga, you should end up with a smaller number.

Application

*How many **milliamperes** is 5 **amperes**?*

To solve these conversion problems, first sort out what value you are converting ***from*** and what value you are converting ***to***. For this example, we're going to convert **from amperes to milliamperes**. We're going **from** a standard value **to** a metric value, so we'll divide the standard value by the power of 10 for milli, 10^{-3}. We should expect to end up with a number larger than 5. How much bigger? Three decimal places bigger!

To divide by a negative exponent, we move the decimal point to the right, and since the exponent is -3, we'll move it three places. You'll need to add some zeroes.

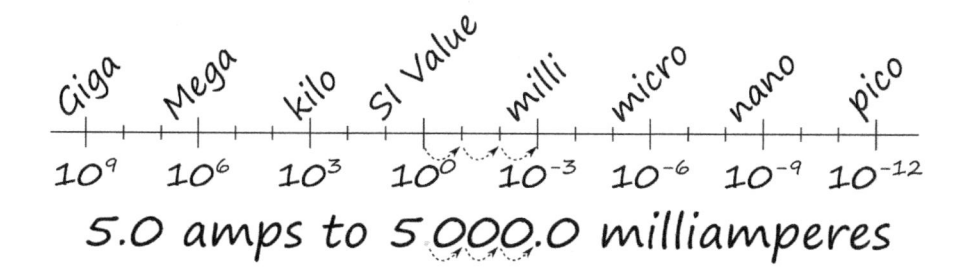

```
  Giga   Mega   kilo  SI Value  milli  micro  nano  pico
   |──────|──────|──────|⌒⌒⌒|──────|──────|──────|
  10⁹   10⁶   10³   10⁰  10⁻³  10⁻⁶  10⁻⁹  10⁻¹²
```
5.0 amps to 5,000.0 milliamperes

Double-check the result. We're going from a "big" unit to a "smaller" unit – there are lots of milliamperes in an ampere. (1,000 of them, to be precise.) We should have a bigger number than we started with, and we do.

How many megavolts is 18,000,000 volts?

In this example, we are converting a smaller value (volts) to a larger value (megavolts). That's the reverse of what we did in Example 1.1. We'll divide the value by the power of 10 for mega, 10^6. Our answer should be a number of megavolts considerably smaller than 18,000,000 volts.

We're moving ***from* volts** (10^1, which equals 1) ***to* megavolts** (10^6.) We'll move the decimal point to the left, this time 6 places, transforming 18,000,000 volts to 18 megavolts.

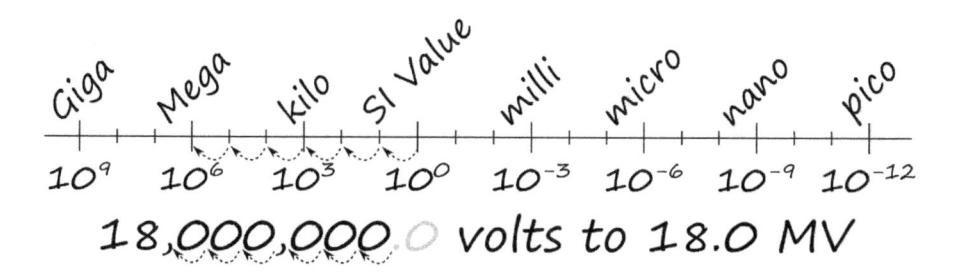

```
  Giga   Mega   kilo  SI Value  milli  micro  nano  pico
   |──────|⌒⌒⌒⌒⌒|──────|──────|──────|──────|──────|
  10⁹   10⁶   10³   10⁰  10⁻³  10⁻⁶  10⁻⁹  10⁻¹²
```
18,000,000 volts to 18.0 MV

Convert 14,313 kilohertz to megahertz.

This is a conversion that you will do a lot when you begin using HF. Some folks – and radios – like to specify HF frequencies in MHz, others in kHz. Our HF radio's frequency read-out is in MHz. Some of the "frequency charts" that show the legal frequencies for different licenses are in kHz. It happens so often, you'll quickly learn to do this conversion automatically in your head.

This one is a little different than our previous examples, since we're starting at a metric value and converting it to another metric value.

Our method doesn't really change, though. We first calculate the difference in powers of 10. This time we're going ***from* kilohertz** (10^3) ***to* megahertz** (10^6), a difference in exponents of 3. We'll move the decimal place of 14313.0 three places to the *left* and get our answer of 14.313 megahertz.

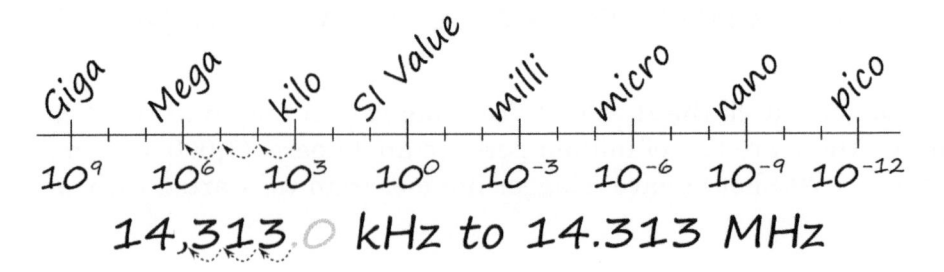

Giga   Mega   kilo   SI Value   milli   micro   nano   pico
10^9 10^6 10^3 10^0 10^{-3} 10^{-6} 10^{-9} 10^{-12}

14,313.0 kHz to 14.313 MHz

Chapter 4

Wavelengths and Frequencies

You'll very often hear hams referring to the such-and-such "meter band", or saying something like, "20 meters opened up and I was talking to Japan!" It's a shorthand way to refer to a particular set of frequencies by their wavelength in meters. Each band has its own set of characteristics, opportunities, and challenges.

So, what is wavelength? A radio signal is created by an oscillating electrical wave, going from zero to maximum positive charge, back to zero, then to maximum negative charge, and finally back to zero. That whole journey is called a cycle. If we graph it, with time going left to right, and charge going from the bottom to the top, the wave looks something like Figure 4.1.

It is always the measurement of one complete cycle of zero to positive, back to zero, to negative, and finally back to zero. (We could also measure from crest to crest; in other words, from a maximum positive point to the next maximum positive point or from a maximum negative point to the next maximum negative point.)

That wave is going to travel at the speed of light, no matter what – no matter what its frequency, no matter what power or type of signal our transmitter is putting out. That means each cycle will occupy a certain amount of space, and that amount of space will be determined by the frequency.

Think of it this way; rather than invisible waves, let's imagine a train track where the trains always travel at exactly the same speed. We'll say

```
C +
h
a
r
g
  -
e
Time
```

|←— One Wavelength —→|

|←— One Wavelength —→|

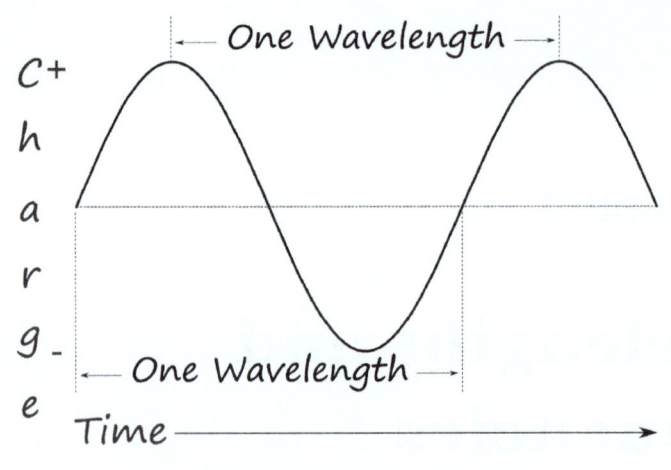

Figure 4.1: One Wavelength = One Cycle

a train is pulling two cars on this track.

Let's say it takes two seconds for those two cars to pass by us. We have a "frequency" of one car per second. (We're ignoring the engine in this example, we're just concerned with the cars.)

Remember, these trains always travel at exactly the same speed. If we knew that speed, we'd know how long each car in the train is. We'll say the speed is always 100 feet per second. One car passed in one second, so that car must be 100 feet long. With a little imagination, we could say these train cars have a "wavelength" of 100 feet. You can see that, so long as the speed remains constant, frequency and wavelength are just different ways of measuring the same thing.

We can write this mathematically as a fraction. Whenever we see the word "per", that means "divide by", so our fraction for the length of a car looks like this:

$$\frac{100 \ feet \ per \ second}{1 \ car \ per \ second} = 100 \ feet \qquad (4.1)$$

We can turn this into a general rule, because that speed always remains the same:

$$Length \ of \ 1 \ car = \frac{100 \ feet \ per \ second}{Number \ of \ cars \ per \ second} \qquad (4.2)$$

That's the formula we would use if we knew how many cars passed by each second and we wanted to know the length of each car.

If we knew the length of the cars but didn't know how many cars would pass by each second, we could use a different fraction.

$$\frac{100 \ feet \ per \ second}{100 \ feet} = 1 \ car \ per \ second \qquad (4.3)$$

And turning that into a general rule, we'd say:

$$Cars \ per \ second = \frac{100 \ feet \ per \ second}{Length \ of \ 1 \ car} \qquad (4.4)$$

We wait by the side of the track for a while, and another train comes along – still traveling at exactly the same speed.

This time, four cars go past us in two seconds. That's two cars per second, twice as many cars as in the first example. We know that each car in this train must be half as long as the cars in that first train!

Let's see if that works with our formula.

$$Length \ per \ car = \frac{100 \ feet \ per \ second}{Number \ of \ cars \ per \ second} = \frac{100}{2} = 50 \ feet \qquad (4.5)$$

23

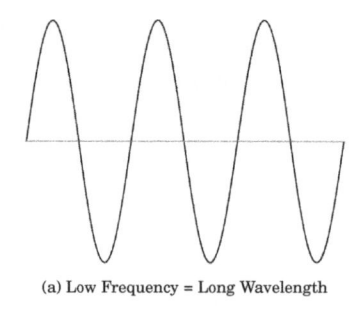

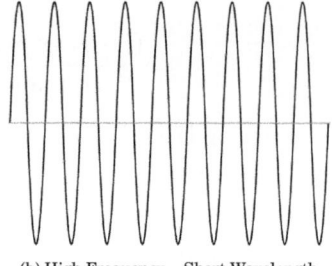

(a) Low Frequency = Long Wavelength (b) High Frequency = Short Wavelength

Figure 4.2: Frequency & Wavelength Relationship

At a frequency of two cars per second, at a speed of 100 feet per second, each car must be 50 feet long.

What if someone told us, "The train that's coming has cars that are 50 feet long?" We could figure out how many cars we'd see in one second with our "Number of cars per second" formula.

$$Cars\ per\ second = \frac{100\ feet\ per\ second}{Length\ per\ car} = \frac{100}{50} = 2\ cars\ per\ second \quad (4.6)$$

The formula says we'll see two cars pass per second; if we know the wavelength, we know the frequency and vice versa.

Now let's get back to radio waves.

The number of cycles per second (Hertz) is the frequency of the signal. It's "how many train cars go by each second."

Because radio waves are moving away from the antenna, we can think of them much the way we might think of waves in the open ocean. If we could somehow see those waves in the air, and get them to hold still, we could measure one with our handy radio wave tape measure and we'd know the wavelength.

Every frequency has a wavelength. The higher the frequency, the shorter the wavelength. See Figure 4.2. Figure 4.2a shows a relatively low frequency signal; three cycles in the time shown on the graph. Figure 4.2b shows nine cycles happening in the same amount of time, with a wavelength one-third that shown in Figure 4.2a.

Why are we so interested in wavelength? For one thing, it's a lot easier to talk about "the 2-meter band" than to say "the range of amateur radio frequencies between 144.000 and 148.000 Megahertz!" More importantly, wavelength affects *everything* about radio, from the size and shape of our antenna to the type of wire required for the feed line between the antenna and transceiver, what sorts of connectors we can use to make those connections, and the propagation results we can expect.

The most popular ham bands cover wavelengths from 160 meters – about 525 feet, or $1\frac{3}{4}$ football fields – to 70 centimeters – about 28 inches. Less popular frequencies go as high as some with wavelengths measured in *millimeters*. That's quite a range of scale.

It's important to remember when we speak of, say, the 2-meter band, that's a range of frequencies, and it's an approximation of the wavelength. In fact, for the 2-meter ham radio band, *none* of the frequencies will come out to precisely 2 meters, but we weren't going to call it the 2.054794 meter band!

Wavelength is a fundamental of ham radio, so if you're not crystal clear on it yet, let us point you toward our *Fast Track* video on the subject on the *Fast Track* web site, fasttrackham.com, in the Teaching Videos section.

We can use simple maths to figure out the wavelength of any frequency. We calculate wavelength by dividing the *speed of light in meters* by the *frequency*. Don't panic. The speed of light is very close to 300,000,000 meters per second. To simplify, we divide 300 by the frequency *in MHz*. The "millions" balance out and the answer comes out in meters.

Here's the formula for calculating the wavelength in meters from a known frequency:

$$Wavelength\ in\ meters = \frac{300}{Frequency\ in\ MHz} \qquad (4.7)$$

To calculate the frequency from the wavelength, the formula is:

$$Frequency\ in\ MHz = \frac{300}{Wavelength\ in\ Meters} \qquad (4.8)$$

A tool for remembering the relationships of frequency, wavelength, and the speed of light, is what we might call the Wavelength Wheel.

Speed of Light (300,000,000 m/s)

300 / λ | f

Wavelength (m)   Frequency (MHz)

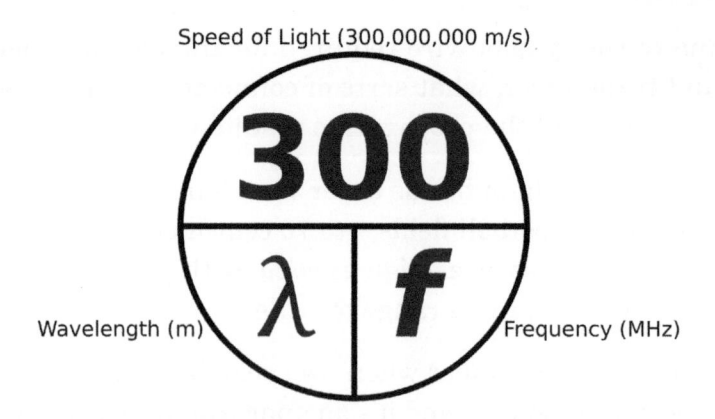

λ is the symbol for *wavelength* and f is the symbol for *frequency*. To use the wheel, you cover the value you don't know. What's left is the formula for finding the unknown. If you don't know the wavelength, you cover the λ. What's left is $\frac{300}{f}$. (Of course, you could also calculate the speed of light by multiplying the frequency by the wavelength, but we already know that answer.)

Application

The first area where these formulas are at least a little helpful is knowing which band is which. If someone mentions the 20-meter band, you'll be able to quickly divide 20 into 300 and know that the 20-meter frequencies are somewhere around 15 MHz. (14.000 to 14.350 MHz, to be more precise.)

Which amateur band contains the frequency 18.110 MHz?

We'll use this formula $\lambda = \frac{300}{f}$ to convert frequency to wavelength:

$$\lambda = \frac{300}{f} = \frac{300}{18.110} = 16.57 \, m \approx 17 \, m \qquad (4.9)$$

At this point, we simply need to know there's no 16-meter amateur band, so we round 16.57 meters up to 17 meters.

What is the approximate value in MHz of frequencies in the 20-meter band?

For this, we use $f = \frac{300}{\lambda}$.

$$f = \frac{300}{\lambda} = \frac{300}{20 \, m} \approx 15 \, MHz \qquad (4.10)$$

A quick look at an amateur radio frequency chart will tell us that 15 MHz is outside the legal range of our 20-meter band, and conversions from frequency to amateur band are nearly always rough estimates. Basically, this tells you that 1.8 MHz is definitely NOT in the 20-meter band.

The other application of these formulas is calculating antenna lengths, a vital skill for most amateurs.

What is the approximate length, in inches, of a half-wave dipole antenna for 147 MHz?

Resonant antennas work best when they are the proper length. That length relates directly to wavelength. If we're building a half-wave antenna for 147 MHz we want to know "how long is one-half of a **wavelength** of a signal at a **frequency of 147 MHz?**"

First we need to calculate the length of one **wavelength** at **147 MHz**.

$$\lambda = \frac{300}{f} = \frac{300}{147\ MHz} = 2.04\ m \qquad (4.11)$$

Now we need to divide that **2.04 m** in half, because we want the length of $\frac{1}{2}$ wavelength.

$$\frac{1}{2}\lambda = \frac{\lambda}{2} = \frac{2.04\ m}{2} = 1.02\ m \qquad (4.12)$$

If you want the answer in inches, here's the formula. There are 39.37 inches in a meter.

$$Inches = Meters \times 39.37 = 1.02\ m \times 39.37 = 40.16\ in \qquad (4.13)$$

You can also calculate in feet directly by substituting 492 for 300 in the wavelength formula.

$$\frac{1}{2}\lambda = \frac{492}{f} = \frac{492}{147\ MHz} = 3.34\ ft \qquad (4.14)$$

You could even go directly to inches by substituting 5,904 (12 × 492) for 300 in the formula.

$$\frac{1}{2}\lambda = \frac{5904}{f} = \frac{5904}{147\ MHz} = 40.16\ in \qquad (4.15)$$

All these formulas are approximations, as you can tell by the slight deviations in the answers. None of them take into account the precise value of the speed of light through a vacuum (299,792,458 meters per second), nor the precise value of the speed of propagation through the antenna, which is slower than 299,792,458 meters per second. In practice, you'll cut your antenna a little longer than any of these values, then trim it using some sort of antenna analyzer. [1]

Finally, if you are taking a ham exam in the US, substitute 468 for 492 in the equations for feet and inches – the 492 number is for ideal antennas in free space while 468 is more accurate for real-world antennas and is the figure used to create US exams.

---

[1]The precise speed of propagation through the antenna can be affected by the design of the antenna, its location, and the material used to make it.

Chapter 5

Decibels – The Comparison Value

We use decibels constantly in amateur radio. It would be a rare occasion indeed that would require you to know the technique of decibel calculation, which looks daunting but is actually very simple. You will, though, be well served by having a general idea of how decibel calculations work and their uses.

We most often use decibels to measure "gain." Put simply, gain is "how much louder or more powerful is this thing than that thing?" dB are what is called a "dimensionless unit", which is a fancy mathematical way of saying they're meaningless until we know the starting point – the "that thing" we're comparing "this thing" to.[1] We might use dB to compare the performance of two amplifiers, or two antennas, or even to compare the performance of different kinds of transmission line that take our signal from the transmitter to the antenna, but dB are *always* about comparison.

Decibels are a *logarithmic* scale, and each 10 dB of gain represents a ten times gain in power. Negative 10 dB (-10 dB) means the power was reduced to a tenth of what it was. (We will reveal the mysteries of logarithms shortly.)

Part of what makes dB so useful is that they can be easily added or subtracted from each other. So if we have something that gives us 47 dB

---

[1]The plural form of decibel is decibels, but the plural form of dB is dB. In this way we sow confusion and thus confound our enemies.

Decibels			
dB	Gain	dB	Loss
3 dB	$Power \times 2$	-3 dB	$\frac{Power}{2}$
6 dB	$Power \times 4$	-6dB	$\frac{Power}{4}$
10dB	$Power \times 10$	-10 dB	$\frac{Power}{10}$

Table 5.1: Decibel Power Gains and Losses

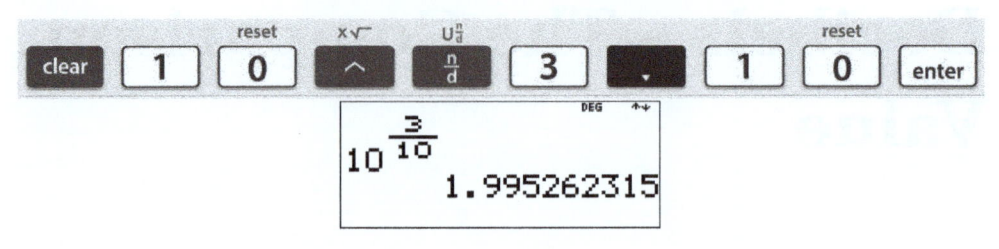

Figure 5.1: Calculator Keystrokes and Screen for Equation 5.1

of gain, and something else that produces 23 dB of loss, we can simply subtract 23 dB from 47 dB

A doubling of power always produces 3 dB (decibels) of gain. Even if you start at 100 dB, 3 dB of gain means the power doubled. 6 dB is two doublings, or a gain of four times. Negative 3 dB (-3 dB) of gain means the power got cut in half.

When you remember 3 db equals a doubling of power, you can estimate any other values you might need. 2 dB is approximately the original value plus 2/3 of the original value.

All you need to remember about dB is Table 5.1

Here's the formula for the power factor:

$$Power\ Factor = 10^{\frac{dB}{10}} \tag{5.1}$$

10 raised to the power of the number of dB divided by 10. If you work that on your scientific calculator, you'll find 3 dB = precisely 1.995262315 times increase in gain. Close enough to 2 times for our purposes! You can see exactly how to do that calculation on your TI-30XSA in Figure 5.1.

To convert a multiple of power – called a power ratio or power factor – into dB, the formula is

30

$$dB = 10 \times log_{10}(Power Factor) \qquad (5.2)$$

That's "10 times the logarithm$_{10}$ of the power factor." A logarithm is sort of the mirror image of an exponent. It represents the power to which the base must be raised to equal a given number. $10^2 = 100$, so the base 10 logarithm of 100 (written as $log_{10}(100)$) is 2. While it is possible to have logarithms that use other bases, by far the most common in our hobby is log_{10}. In fact, log_{10} is so common that the **only** whole-number logarithm most calculators will compute directly is log_{10}.[2]

There are 10's in both of those formulas because a decibel is 1/10th of a bell, so we're converting from bells (a value no one in ham radio uses for anything) to decibels. We can try out our formula with a power factor of 2, representing a doubling of power.

$$dB = 10 \times \log_{10}(2) = 3.010299\ dB \qquad (5.3)$$

And, sure enough, a doubling of power is the equivalent of about 3 dB of gain.

Most calculators, including the TI-30XS, only do "log$_{10}$" so if you try entering that calculation above in your scientific calculator, you'll just enter

$$10 \times log(Power\ Factor) \qquad (5.4)$$

You don't need all that jazzy maths to pass any ham exam we have ever seen, nor do you need such precise answers in most practical applications. (Michael has worked with dB since shortly after lightning was invented and has yet to need the sort of precision these calculations provide, but it is good to understand the maths behind dB.)

---

[2]Many scientific calculators, including the TI-30XS, will calculate *natural* logarithms, which have a base of the number e. e is approximately 2.718 and is very useful in advanced maths, including calculus. If, for some reason, you find yourself often needing logarithms of bases other than 10 or e, some Casio calculators offer the ability to calculate logs in any base.

Application

What is the approximate amount of change, measured in decibels, of a change in power from 750 watts to 1500 watts?

To calculate this, we need to know the power factor (the ratio) between 750 watts and 1500 watts. To work these problems, we always divide the **starting value** *into* the **final value**. In this case, we are starting with **750 watts** and our final value is **1500 watts** because we're changing the power from 750 watts to 1500 watts. (As you'll see shortly, if we flipped those numbers so that 750 was on the top of the fraction, the answer would be the power factor of going from 1500 to 750 – so it's important to figure out the "from" and "to" numbers.)

$$Power\ Factor = \frac{Final\ Value}{Starting\ Value} = \frac{1500\ W}{750\ W} = Power\ Factor\ of\ 2 \quad (5.5)$$

We *could* calculate the exact dB answer with the same calculation we did as an example, which we know comes out to 3.010299 dB. However, since you've memorized the fact that a **two-times increase in power** equals **3 dB** of gain, you know that's the answer. *Every* question we have seen on any amateur radio exam asks about "approximate dB", never exact figures like 3.010299.

What is the approximate amount of change, measured in decibels, of a decrease in power from 1500 watts to 750 watts?

Obviously, this is the reverse of the previous example. To calculate our answer, we'll again divide the **starting value** into the **final value**.

$$Power\ Factor = \frac{Final\ Value}{Starting\ Value} = \frac{750\ W}{1500\ W} = Power\ Factor\ of\ 0.5 \quad (5.6)$$

The power has been cut in half, and we know from the chart that cutting power in half equals a -3 dB change, which we'd call "-3 dB of gain" or "3 dB of loss." Just for grins, let's do it the hard way:

$$dB = 10 \times log_{10}(Power\ Factor) = 10 \times log_{10}(0.5) = -3.010299\ dB \quad (5.7)$$

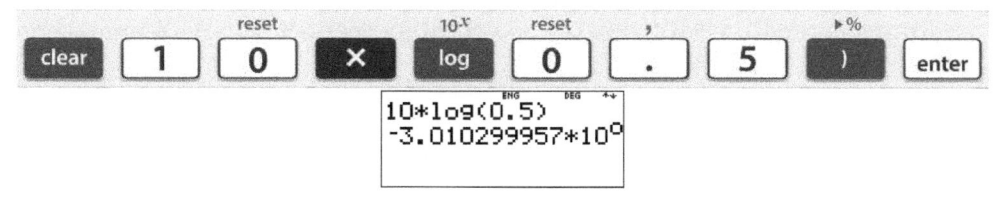

Figure 5.2: Keystrokes for Equation 5.7

Figure 5.2 shows the keystrokes for the TI-30XS and, sure enough, cutting the power in half reduces the signal strength by $3\,dB$.

If we increase the power of a 100-watt amplifier by 10 dB, what will its new power be?

A 10 dB increase in power is a 10x increase in power. We're increasing the power of a **100 watt** amplifier by **10 dB**. All we need to do is multiply **100 watts** by 10, which you can probably do in your head, but we'll write it out anyway.

$$100\ W \times 10 = 1000\ W \tag{5.8}$$

Let's see how close the formal maths says that power factor of 10 is:

$$Power\ Factor = 10^{\frac{dB}{10}} = 10^{\frac{10}{10}} = Power\ Factor\ of\ 10 \tag{5.9}$$

Unlike other dB values, 10 dB is an exact number; *exactly* a 10x increase or decrease in power.

Fill in the blank: Handheld transceiver A has an output power of 5 watts. Transceiver B has an output power of 8 watts. This means Transceiver B has _____ dB of gain compared to transceiver A.

We start by calculating the power factor of the two powers; in other words, the ratio between them. We can treat this just the way we treated previous power factor problems, with a starting value (5 watts) and a final value (8 watts.)

$$Power\ Factor = \frac{FinalValue}{StartingValue} = \frac{8\ W}{5\ W} = Power Factor\ of\ 1.6 \tag{5.10}$$

33

Transceiver B is 1.6 times as powerful as transceiver A. What's that in dB?

$$dB = 10_{log}(Power Factor) = 10_{log}(1.6) = 2.04 \ dB \qquad (5.11)$$

Transceiver B has **2.04 dB** of gain compared to transceiver A. (In other words, not much!)

Chapter 6

Ohm's Law & Joule's Law

The cornerstone for all the maths that creates the science of electronics is Ohm's Law. Georg Simon Ohm's Law describes the relationship among voltage, current and resistance in an electrical circuit. Those three values are always interconnected. If you increase the voltage into the same resistance, more current will flow. Keep the voltage the same and reduce the resistance, more current flows again.

The Law says: Voltage equals Current multiplied by Resistance.

$$E = I \times R \qquad (6.1)$$

E = Voltage in volts
I = Current in amperes
R = Resistance in ohms

Why in the world would they make "**E**" stand for voltage and "**I**" stand for current? Well, when Georg Simon Ohm published his Law in 1827, the volt had not yet been named for Volta – that wouldn't happen until 1893. In Ohm's time, voltage was "electromotive force" or "**E**" for short.[1]

The ampere had not been named for Andre-Marie Ampére yet, either. Ampére called electric current "intensité", French for intensity. Thus, the "I" in our formula.

Of course, Ohm didn't name the unit of resistance after himself, either.

When I (Michael) was first learning this, it struck me as a wildly unlikely coincidence that all those values worked out so neatly. In fact, it was no coincidence at all. The volt, ampere, and ohm were defined to match *because* of Ohm's Law.

---
[1]Being German, Georg probably called it *elektromotorische Kraft*. Still "E."

So, we have had it handed down to us as $E = I \times R$ and that is the standard on *some* amateur radio exams and reference sources. Others use V in place of E and we'll be following that example.

With a tiny little bit of algebra, we can turn $V = I \times R$ into

$$I = \frac{V}{R} \tag{6.2}$$

or

$$R = \frac{V}{I} \tag{6.3}$$

Those formulas let us figure out either **V**, **I**, or **R** so long as we know the two other values.

Here's an easy way to remember all the variations. We call it Ohm's Pie.

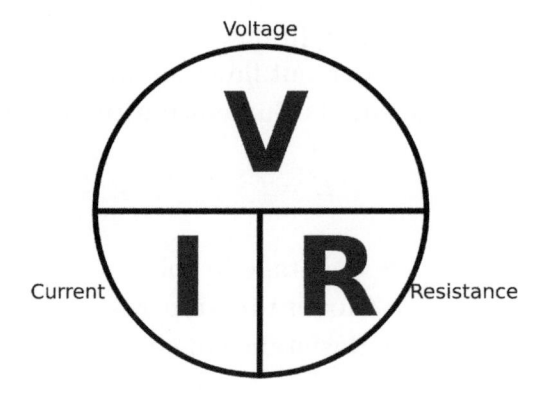

Ohm's Pie works just like the Wavelength Wheel from Chapter 4. To use it, you just cover up the value you don't know, and you'll be left with the formula to figure it out. If you want to know the current in a circuit, you cover up the I and you're left with V/R. Divide the volts by the resistance and you'll get the current. Need to know the voltage? Cover up the V and what's left is $I \times R$.[2]

You can't take a copy of Ohm's Pie into amateur license exams, but you can certainly draw an Ohm's Pie chart on your scratch paper while you take the test – that's perfectly legal. As you'll see on the questions below, it can be *very* helpful to have that chart.

---

[2]In fact, you can make a similar wheel out of any equation that takes the form $x = y \times z$. One author of this book may have used this trick to (barely...) survive high school algebra.

The other formula you'll need for this section is the formula for power – the watts used in a circuit. That's known as Joule's Law. The formula for watts is:

$$P = I \times V$$

Power equals voltage times current. Again, there's a simple chart we'll call the Power Pie that will let you calculate wattage when you know two of the values, and it works the same way as Ohm's Pie.

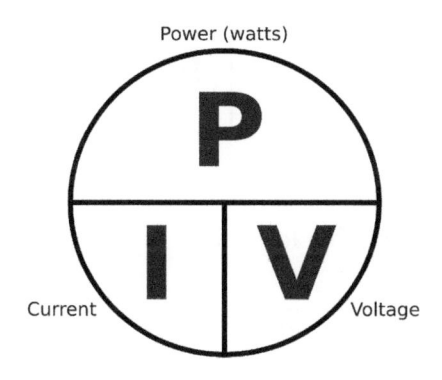

If you want to know the voltage of a circuit carrying 10 amps and using 100 watts, you cover up the "V" and see that

$$V = \frac{P}{I}$$

Divide 100 watts by 10 amps and get 10 volts as the answer.

So long as you correctly identify what value the question is asking for, the rest is simple maths. Sometimes, in the real world, an Ohm's Law related problem will involve two steps; "How many watts will be used by a circuit with 240 volts across 100 ohms of resistance?", for instance. In that case, you'd need to calculate the current in the circuit, then multiply the current by the voltage for the result. (There are formulas for solving power problems in one step – $P = \frac{V^2}{R}$ or $P = I^2 \times R$ – but we find it easier to just remember the pies and do it in two steps.)

Application

What is the voltage across a circuit with 4 amperes of current and 2 ohms of resistance?

The first step in any of these problems is identifying the value the question wants. In this case, it's **V**, the voltage across the circuit.

37

Next, we want to know which known values they have given us. This question gives us the current, I, and the resistance, **R**. Consult your Ohm's Pie and you can see you have everything you need to solve this one:

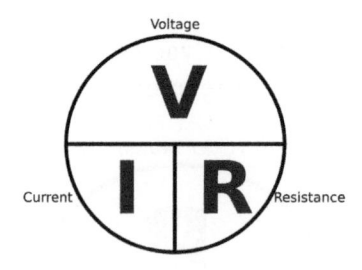

Cover up the V and you're left with $I \times R$.

$$V = I \times R = 4 \times 2 = 8 \text{ volts}$$

What is the current through a circuit with 240 volts across it and 150 ohms of resistance?

This question gives us the voltage across and resistance in the circuit and asks for the current. A look at Ohm's Pie tells us the formula.

$$I = \frac{V}{R} = \frac{240}{150} = 1.6 \text{ amperes}$$

What is the total resistance in a circuit that carries 15 amperes of current when 120 volts is applied across the circuit?

This question asks us for the resistance in the circuit and gives us the voltage and current. Once again, we have everything we need to solve the problem in one step.

Covering the unknown, the R, in Ohm's Pie leaves the formula:

$$R = \frac{V}{I} = \frac{120 \text{ V}}{15 \text{ A}} = 8 \, \Omega$$

How many watts are being consumed in a circuit carrying 15 amperes at 120 volts?

This question asks about watts, the measure of power. There's no "P" for power on our Ohm's Pie, but there is on the Power Pie.

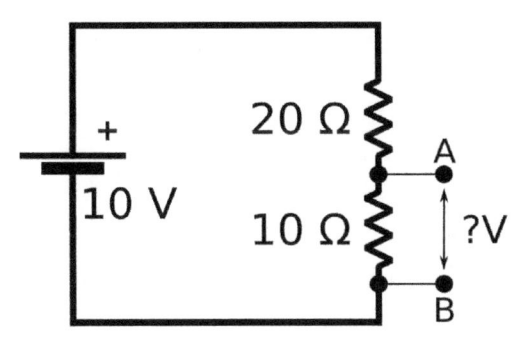

Figure 6.1: Voltage Divider

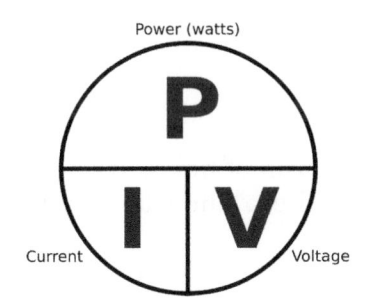

We're given the current, **I**, and the voltage, **V**, so we can multiply those two together to get the watts, **P**.

Consider the circuit shown in Figure 6.1. This circuit produces a voltage at the output, which is between points A and B. What is that voltage?

The circuit shown in Figure 6.1 is called a *voltage divider* and it does just what the name says. This is a very common circuit, used to produce a voltage lower than the supply voltage for some component that needs to be supplied with that voltage.

This introduces the concept of *voltage drop*, which we calculate using Ohm's Law. The voltage drop across a component is just another name for the voltage across a component. Remember, voltage is always a *difference* of electromotive force.

There are a couple of ways to solve this problem. If we measure the voltage across both resistors at once, it will equal the source voltage; it must, since it would electrically be the same as measuring across the terminals of that battery. There's a formal statement of this, known as Kirchoff's Law of Voltage. That Law tells us that the total voltage drop across the circuit *must* equal the source voltage. It follows that the voltage drop of each resistor must be proportionate to the fraction of the total resistance represented by that resistor. In other words, if the resistor accounts for half the total resistance, it accounts for half the voltage drop.

In this case, we have different values of resistors, so we can't say each accounts for one-half. Let's calculate.

$$R_{Total} = R_1 + R_2 = 20\,\Omega + 10\,\Omega = 30\,\Omega \tag{6.4}$$

Now we can calculate the current in the circuit.

$$I = \frac{V}{R} = \frac{10}{30} = .33\,A \tag{6.5}$$

Now we can calculate the voltage drop across each resistor. We'll call the 20 Ω resistor R_1 and the 10 Ω one R_2.

$$V_{R1} = I \times R_1 \approx 0.33\,A \times 20\,\Omega \approx 6.7\,V \tag{6.6}$$

R_1 gives us a voltage drop of 6.6 V. It is easy to guess what voltage drop occurs across R_2, but here's how to calculate it.

$$V_{R2} = I \times R_2 = 0.33\,A \times 10\,\Omega \approx 3.3\,V \tag{6.7}$$

The voltage between points A and B is 3.33 volts.

There's an elegant one-step formula for solving voltage dividers, thanks to our new best friend, Al Jabra.

$$V_{out} = \frac{V_s \times R_2}{R_1 + R_2} = \frac{10\,V \times 10\,\Omega}{20\,\Omega + 10\,\Omega} \approx 3.33\,V \tag{6.8}$$

Truly understanding Ohm's Law and its cousin Joule's Law of Power are critical to your understanding of electricity and electronics. The applications can be as obvious as finding the answer to "why did the circuit breaker trip when I turned on the toaster, the blender, and the coffeemaker" or as subtle as understanding why lightning protection systems for outdoor antennas work best when all ground rods are electrically bonded to each other.

Chapter 7

Series & Parallel Calculations

We think most people have what we'd call an "exit point" from mathematics. For some it might be long division or Algebra I. For others it is some branch of mathematics way up there at the doctoral level, such as Mathematical Theory Analysis.

We haven't done a scientific survey, but from lots of conversations with hams, the topic of this chapter seems to be a common exit point for electronics students and would-be ham radio operators. We can understand why; at first glance, it seems like a *lot* of formulas, and half of those formulas look more than a little daunting. Really, though, it's only two formulas, and one of them is primary school maths, if that.

A series circuit is a circuit with only one path for the electricity to follow. A parallel circuit has more than one path. It's possible to construct a series/parallel circuit, but those are analyzed using the formulas for series and parallel circuits then combining the results. The total resistance of the series/parallel circuit shown in Figure 7.1 can be analyzed in three steps.

We calculate the resistance of the parallel section, then calculate the resistance of the series section, and finally combine the two as though the parallel section was a single resistor in series with the other two

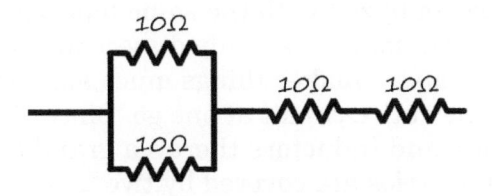

Figure 7.1: Series-Parallel Circuit

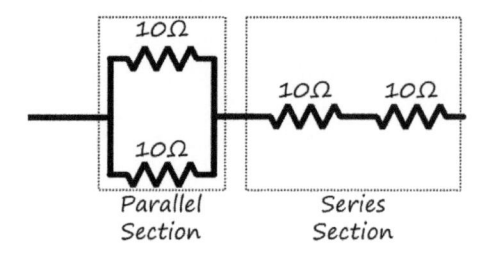

Figure 7.2: Analyzing a Series-Parallel Circuit

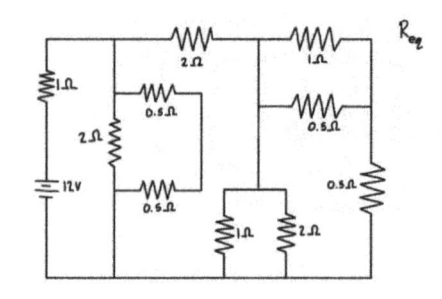

Figure 7.3: Ridiculous Resistance Problem

resistors.

We'll show you the full formulas shortly, but when two resistors of equal value are connected in parallel, the resulting resistance is one-half the value of one of the resistors, so the parallel part of the circuit accounts for 5 ohms (Ω) of resistance. To many people, this seems paradoxical; how can we reduce resistance by adding a resistor? It may seem less puzzling if you think of each additional resistor as an additional pathway of *conductance* (G), which is the reciprocal of resistance. ($G = \frac{1}{R}$.)

The resistors in series simply add together; we have not added any paths of conductance, we have only added resistance. Those two resistors account for 20 Ω of resistance.

Add the parallel resistance of 5 Ω to the series resistance of 20 Ω and the total works out to 25 Ω.

Electrical engineering professors are fond of torturing their students by demanding they analyze circuits such as you see in Figure 7.3, yet even that monstrosity – for which we can imagine no useful purpose whatsoever – can be analyzed with the same tools as Figure 7.1.[1]

The fact that we can mathematically dissect then reassemble circuits in the way described above makes things much simpler for us than if we needed to analyze the entire circuit at one go. Since we'll be dealing with resistors, capacitors, and inductors there's a grand total of six possible scenarios. All six scenarios are covered by two "generic" formulas.

---

[1]That whole thing equals a 1.521 ohm resistor.

Here's the first one:

$$Total = Value_1 + Value_2 \qquad (7.1)$$

That's the one to use for *series* resistors and inductors, and for *parallel* capacitors. You can add as many $Value_n$'s as needed.

Here's the second:

$$Total = \frac{1}{\frac{1}{Value_1} + \frac{1}{Value_2}} \qquad (7.2)$$

That's for *parallel* resistors and inductors, and for *series* capacitors. Again, you can add as many $\frac{1}{Value_n}$'s as there are components.

What those formulas mean is that if you add another resistor or inductor to another resistor or inductor in series, the total resistance or inductance will go up. If you add a resistor or inductor in parallel, the total resistance or inductance will go down.

Capacitors – as they often do! – act in exactly the opposite way. If you add a capacitor to another capacitor in parallel, the total capacitance goes up. The reciprocal of capacitance is *elastance*, measured in *darafs* or just F^{-1}.

Elastance is a capacitor's opposition to accepting an electrical charge. When we hook up capacitors in series, we're adding more opposition to accepting a charge. In a parallel circuit with two equal capacitors, the current "sees" half the elastance it would see if the two capacitors were in parallel.

We can also view this as a result of the physics of capacitance. The capacity of a capacitor depends on several things, but one is the size of the plates. To electricity, a capacitor in parallel with another just looks like one capacitor with bigger plates. So, this parallel circuit...

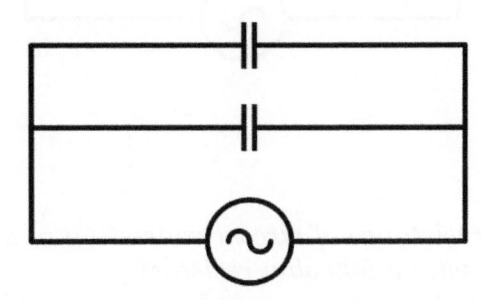

...is, electrically speaking, the same as this:

43

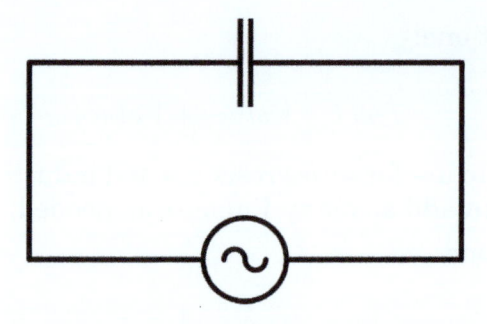

Another factor that affects a capacitor's capacity is the thickness of the dielectric material between the plates. A thinner dielectric increases capacitance. To electricity, two capacitors in series look like a single capacitor with a very thick dielectric.

This series circuit...

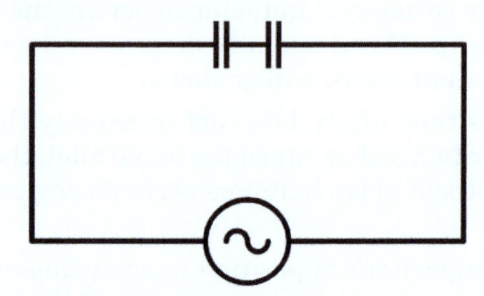

...is the electrical equivalent of this:

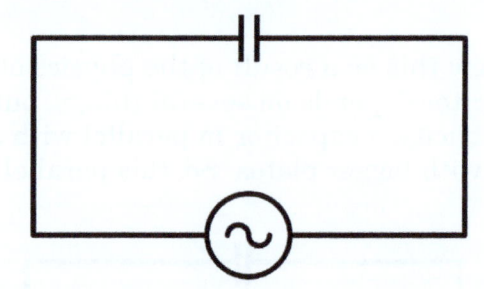

Application

What is the total resistance of circuit consisting of a 100-ohm resistor connected in series with a 500 ohm resistor?

Resistors and inductors connected in series add together. In the addition formula, we'll substitute R (resistance) for *Value* in the generic formula above.

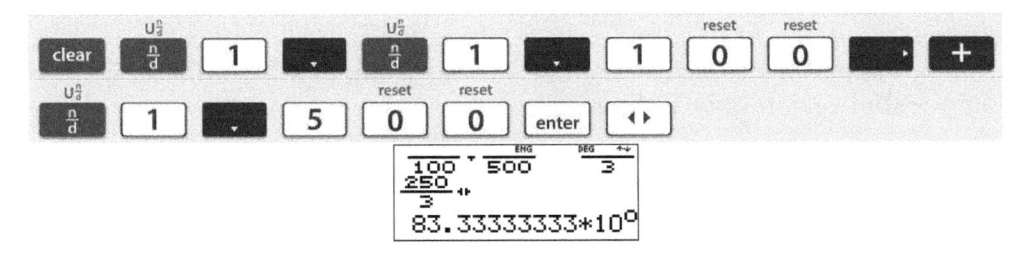

Figure 7.4: Keystrokes for Equation 7.4

$$Total = R_1 + R_2 = 100\,\Omega + 500\,\Omega = 600\,\Omega \tag{7.3}$$

If, instead of resistors, the question was about inductors in series, we'd use exactly the same formula.

What is the total inductance of a circuit consisting of a 100 millihenry inductor connected in parallel with a 500 millihenry inductor?

Now we have two inductors in parallel, so we need the other generic formula. We'll plug in L (inductance) for *Value*. We don't even need to convert any values, because everything's in millihenries already.

$$Total = \frac{1}{\frac{1}{L_1} + \frac{1}{L_2}} = \frac{1}{\frac{1}{100\,mH} + \frac{1}{500\,mH}} = 83.3\,mH \tag{7.4}$$

Here's a place where the TI-30XS really shines. You can enter the calculation just as you see it above. See Figure 7.4.

As with resistance, adding parallel inductors adds the reciprocal of inductance, which is *susceptance*, B. When we look at what the formula for parallel resistance or inductance does we see it totals up the reciprocals of the R or L, then takes the reciprocal of those values, converting them back to R or L.

If you're doing these by hand, this is one where writing out each step just the way you see it done above really helps keep things clear through all the steps needed to get to the solution. (I guess this was why my algebra teacher kept saying, "Show your work!" However, that would have involved actually *doing* some work...)

What is the total capacitance of a circuit consisting of a 100-microfarad capacitor connected in parallel with a 500-microfarad capacitor?

Capacitors in parallel add together just like resistors and inductors in series.

$$Total = C_1 + C_2 = 100\,\mu F + 500\,\mu F = 600\,\mu F \tag{7.5}$$

What is the total capacitance of a circuit consisting of a 100 millifarad capacitor connected in series with a 500 millifarad capacitor?

For capacitors in series, we use the same generic formula as we use for parallel resistors or inductors.

$$Total = \frac{1}{\frac{1}{C_1}+\frac{1}{C_2}} = \frac{1}{\frac{1}{100\,mF}+\frac{1}{500\,mF}} = 83.3\,mF \qquad (7.6)$$

See above for the keystrokes – they're the same as for Example 5.2. The whole key to these is to take them step-by-step. You'll be doing three divisions – two to turn the fractions in the denominator into decimals, then the final one to get the final answer.

A Shortcut!

Remember, if a problem involves *equal value* resistors in parallel, simply divide the value of one component by the number of components. The same trick works for equal value inductors in *parallel* or equal value capacitors in *series*, . For instance, for three 100-ohm inductors in parallel, divide 100 by 3 for an answer of 33.3 henries.

Just to prove it works, we'll do it the long way for three 100-ohm resistors.

$$Total = \frac{1}{\frac{1}{R_1}+\frac{1}{R_2}} + \frac{1}{\frac{1}{R_3}} = \frac{1}{\frac{1}{100\,\Omega}+\frac{1}{100\,\Omega}+\frac{1}{100\,\Omega}} = 33.3\,\Omega \qquad (7.7)$$

Yes! It works!

Chapter 8

Peak-to-Peak and RMS Voltage

AC voltage presents us with a problem when we want to use Ohm's Law with it. The problem is that the voltage is constantly varying, as is the current.

If we want to run a calculation of amperage using Ohm's Law, what value do we plug in for the voltage? 120? 0? Split the difference at 60? Well, the answer is none of the above.

What we need is called the RMS voltage. The RMS voltage is the value that produces the same power dissipation in a resistor as a DC voltage of the same value. RMS stands for Root Mean Squared. Technically, it's the square root of the square of an average of all the values in a half-cycle of the sine wave. It sounds like something that might take a long time to calculate, but, in fact, a simple formula does the job.

$$V_{RMS} = \frac{V_{Peak}}{\sqrt{2}} \qquad (8.1)$$

The RMS voltage of a sine wave equals the peak voltage divided by the square root of 2. For that waveform above, with a peak voltage of 100 V, the RMS voltage is 70.71 V. If we put the voltage represented by that sine wave across a resistor, we'd get the same (average) current flow as if we put 70.71 V of DC across it. That RMS voltage is the number we'll use for any Ohm's Law or Joule's Law calculation.

The square root of 2 is about 1.41, but we're going to use its reciprocal, 0.707. Why? It's easy to remember – just think of that ancient old Steve Miller Band song, *Jet Airliner* – he was going to "get on a 707!" We only need that one number to convert from RMS to peak or from peak to RMS.

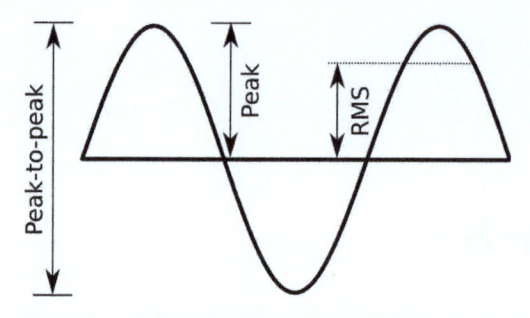

Figure 8.1: Peak-To-Peak, Peak, and RMS Voltages

To convert peak voltage to RMS, multiply the peak voltage by 0.707. To convert RMS to peak voltage, divide the RMS voltage by 0.707.

The other distinction you need for this section is between peak voltage and peak-to-peak voltage. In the waveform above, the peak-to-peak voltage is 200 V. The peak voltage is 100 V. To convert to RMS, we *always* use peak voltage, which is always one-half of peak-to-peak voltage – at least for purposes of ham exams.

Application

What is the RMS voltage of a sine wave with a peak value of 340 volts?
For this example, we're converting from peak voltage to RMS voltage.

$$V_{RMS} = V_{Peak} \times 0.707 = 340\,V \times 0.707 = 240.38\,V_{RMS} \quad (8.2)$$

What is peak-to-peak voltage of a sine wave with an RMS voltage of 1500 volts?

This is a two-step conversion; from RMS to peak voltage, then from peak voltage to peak-to-peak.

$$V_{Peak} = \frac{V_{RMS}}{0.707} = \frac{1500\,V_{RMS}}{0.707} = 2121.64\,V_{Peak} \quad (8.3)$$

We still need to turn that peak voltage, $2121.64\,V_{Peak}$, into a peak-to-peak value.

$$V_{Peak-to-peak} = V_{Peak} \times 2 = 2121.64\,V_{Peak} \times 2 = 4243.28\,V_{Peak-to-peak}$$
$$(8.4)$$

What is the output PEP (Peak Envelope Power) from an RF power amplifier if an oscilloscope measures 550 volts peak-to-peak across a 50-ohm dummy load connected to the transmitter output?

This one requires a conversion from peak-to-peak voltage to peak voltage; conversion from peak voltage to RMS voltage; an Ohm's Law calculation; then, a Joule's law calculation to get the PEP in watts. We want to end up with an answer in watts. First, the conversion from peak-to-peak.

$$V_{Peak} = \frac{V_{Peak-to-peak}}{2} = \frac{550\,V}{2} = 275\,V_{Peak} \tag{8.5}$$

Now we'll convert 275 V **peak voltage** to **RMS voltage**.

$$V_{RMS} = V_{Peak} \times 0.707 = 275\,V \times 0.707 = 194.43\,V_{RMS} \tag{8.6}$$

For convenience, we've rounded that answer to 194.43 volts. That's plenty of precision for any of the ham exams. Now we can calculate the **current** through that 50-ohm dummy load (which is just a big resistor.) It's a straight Ohm's Law problem.

$$I = \frac{V}{R} = \frac{194.43\,V_{RMS}}{50\,\Omega} = 3.89\,A \tag{8.7}$$

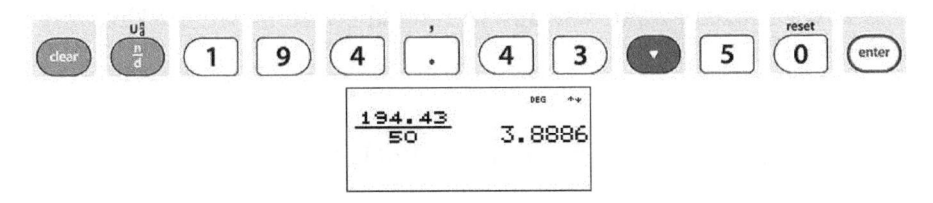

Figure 8.2: Keystrokes for Equation 8.7

We have our RMS voltage and our current, so now we can calculate the PEP output of this amplifier in watts.

$$P_{PEP} = V \times I = 194.43\,V_{RMS} \times 3.89\,A = 756.33\,W \tag{8.8}$$

What is the RMS voltage across a 500-turn secondary winding in a transformer if the 3000-turn primary is connected to 120 VAC?

In most cases, if you are given an AC voltage value with no qualifier, such as "peak-to-peak", it's safe to assume that the voltage given is an RMS voltage. In fact, that's common usage even outside the ham world. Your household voltage is 120 or 220 VAC (volts of AC) RMS.

It's also safe to assume that any component listed in a ham exam question is what's called an *ideal component*. In this case, this is an ideal transformer, one that transforms voltages in precise proportion to the ratio of the turns on the primary and secondary windings. If the primary is 100 windings and the secondary is 10 windings, that transformer will produce exactly 1 volt across the secondary if 10 volts is applied to the primary. In the real world, transformers are subject to internal losses, but this is the ideal world.

For this problem, then, we just need to know the ratio of the turns in the windings and apply that to the voltage across the primary to know the voltage across the secondary.

Here's how the whole formula looks:

$$V_{secondary} = \frac{N_{Secondary}}{N_{Primary}} \times V_{Primary} \qquad (8.9)$$

What we're doing is multiplying the voltage in the primary winding by the ratio of the number of turns (N) in the **secondary** and **primary** windings. Let's take it a step at a time. First, we'll calculate the turns ratio.

$$TurnsRatio = \frac{N_{Secondary}}{N_{Primary}} = \frac{500\,turns}{3000\,turns} = 0.17 \qquad (8.10)$$

Then we multiply the voltage across the primary, 120 VAC, by the turns ratio, 0.17.

$$V_{secondary} = V_{primary} \times Turns\,Ratio = 120\,VAC \times 0.17 \approx 20\,VAC \qquad (8.11)$$

(If you used 0.17 instead of the real ratio of 0.1666667, you got 20.4 VAC. Close enough!)

Chapter 9

Receiver Performance Characteristics

The advertising for modern amateur radio transceivers lists many, many specifications. by way of example, the IC-7300 lists 33 specifications for its receiver section. The transmitter section lists ten.

Happily, most of these specifications are reasonably easy to understand. *Sensitivity* is the ability of a receiver to detect a faint signal. Lower values are better, so $-65\,dB$ would be better than $-64\,dB$. *Selectivity* is the ability of a receiver to pick out one signal from a cluster of signals near the same frequency. That's a little more complicated to measure than sensitivity and there are different standards, but a common one is ACS; adjacent channel sensitivity. ACS is, essentially, "how loud can an adjacent station be before I hear it where it shouldn't be?" Higher numbers are better.

Then there's this puzzling specification; image response. Image response is undesirable; In practice it causes false signals to appear where none should be. The "response" is caused by mixing of two signals in the receiver's Intermediate Frequency section when the numerical distance of one station's frequency from another has the bad luck to relate to the desired frequency in such a way that a "phantom signal" is created in the radio's Intermediate Frequency or IF section.

Image response is what one might call a "special case" of *intermodulation distortion*, which we will cover in Chapter 10. You could even consider it another form of *heterodyning*. All three involve mixing frequencies. When we don't like the result, we call it intermodulation distortion or image response. When we intended the result, as when a receiver's IF section converts an incoming signal to the receiver's intermediate frequency, we call it heterodyning. The mechanisms of all three are identical; mix two frequencies and you get some new frequencies that are the sum and difference of the two original frequencies. (You are also left with the two original frequencies.)

The IF section of a receiver mixes the received signal with a signal created in a part of the receiver's VFO section called the *local oscillator*. The signal created by the local oscillator is offset from the received signal by a set amount. That offset remains the same no matter what frequency the radio is tuned to. The difference between the received signal's frequency and the frequency from the VFO is the receiver's intermediate frequency. In older radios, that frequency was often 455 kHz. Modern radios use higher frequencies, but the principles are the same.

Here's an example. To make the numbers simple, we'll say we're receiving a signal at 15.000 MHz and that our receiver's local oscillator frequency is 500 kHz. That means that when we tune to 15 MHz, the VFO creates a frequency – just a plain old sine wave – at 15.500 MHz. When we mix the received signal with that signal from the VFO, we end up with a copy of the original signal at 500 kHz and another at 15.500 MHz.

$$f_{VFO} - f_{Rcvd\,Signal} = 15.500\,MHz - 15.000\,MHz = \mathbf{500\,kHz} \qquad (9.1)$$

$$f_{VFO} + f_{Rcvd\,Signal} = 15.000\,MHz + 500\,kHz = \mathbf{15.500\,MHz} \qquad (9.2)$$

Since the object of having an IF section is to lower the received signal to a standard and more manageable frequency, we'll filter out the 15.500 MHz signal and the original frequency and send that 500 kHz copy of the original signal to the next stage of the receiver.

That is a description of a *superheterodyne* receiver. Edwin Armstrong invented it to allow the use of higher frequencies than were then in use and it has survived up to the present day. Put succinctly, it is a brilliant invention. Like most brilliant inventions, though, it has an Achille's heel.

The Achille's heel isn't even in the IF section; it's that filter before the IF section that selects what station we're hearing.

What if a station sneaks past that filter – because there are no perfect filters – and that station just happens to be at 16.000 MHz?

$$f_{Rcvd\,Signal} - f_{VFO} = 16.000\,MHz - 15.500\,MHz = \textbf{500 kHz} \qquad (9.3)$$

Houston, we have a problem! That signal 1 MHz away from the signal we are trying to hear is getting mixed with the local oscillator from the VFO and now is at 500 kHz; indistinguishable from the desired signal.

That is image response interference and it will happen any time a signal that is 2× the local oscillator frequency gets past the RF filter. The solution most analog radios use now is simply to raise the local oscillator frequency. That pushes the frequency of the potentially interfering station farther away from the passband of the RF filter.

Application

What transmit frequency might generate an image response signal in a receiver tuned to 14.300 MHz and which uses a 455 kHz IF frequency?
When two frequencies mix, we don't just get one frequency. Put simply, we get two. (We actually get more, but the rest are so weak and far away, we can ignore them, at least for the moment.) One is the sum of the two frequencies; the other is the difference.

In a superheterodyne receiver, we mix the desired signal with a local oscillator signal and that creates a copy of the original signal at a lower frequency, called the Intermediate Frequency, or IF.

The IF stage filters out the original carrier, the local oscillator, and the sum frequency. All that's left is the Intermediate Frequency, and that gets sent along to the rest of the receiver. So far, so good.

Let's do the maths. In this problem, we're tuned to a station at 14.300 MHz, and there's another station at some unknown frequency that is interfering with the one we want to hear. The type of interference is "an image response signal."

Our receiver uses a 455 kHz Intermediate Frequency, and we'll assume it uses what's called "high side injection", meaning the Intermediate

53

Frequency is 455 kHz above the desired station's frequency. (The answer comes out different if we assume low side injection, but that answer isn't one of the possible answers.)

So, we have a 14.300 MHz signal, and another signal from the local oscillator at 14.755 MHz, which is 14.300 MHz plus 455 kHz.

That interference must be coming from a station on a frequency that relates to the desired signal and to the local oscillator frequency, which we'll call f_{local}. We're told it is an *image response*, so we're looking for a difference between two frequencies that gives an answer equal to the intermediate frequency of 455 kHz.

Let's put it in formula form.

$$455\,kHz = f_{unknown} - f_{local} \tag{9.4}$$

or

$$455\,kHz = f_{local} - f_{unknown} \tag{9.5}$$

Important first step: Get everything into the same units. Adding and subtracting kHz and MHz will just create precisely what we do not want, which is confusion.

455 kHz = 455,000 Hz. We'll divide by 10^6, or 1,000,000 to convert Hz to MHz.

$$\frac{455,000}{10^6} = 0.455\,MHz \tag{9.6}$$

We know the local oscillator frequency is 14.755 MHz, so let's plug that in.

$$0.455\,MHz = f_{unknown} - 14.755\,MHz \tag{9.7}$$

Let's do this one by adding 14.755 MHz to both sides.

$$0.455\,MHz + 14.755\,MHz = f_{unknown} \tag{9.8}$$

and...

$$0.455\,MHz + 14.755\,MHz = 15.210\,MHz \tag{9.9}$$

What if we had used the second formula? Let's see.

$$0.455\,MHz = f_{local} - f_{unknown} \tag{9.10}$$

Again, we know the VFO frequency, that's still 14.755 MHz.

$$0.455\,MHz = 14.755\,MHz - f_{unknown} \tag{9.11}$$

Since we have a negative $F_{unknown}$ we'll add an $F_{unknown}$ to each side, cancelling out the one on the right.

$$0.455\,MHz + f_{unknown} = 14.755\,MHz - f_{unknown} + f_{unknown} \tag{9.12}$$

We delete one $f_{unknown}$ on either side....

$$0.455\,MHz + f_{unknown} = 14.755\,MHz \tag{9.13}$$

Then we'll subtract 455 kHz from each side ....

$$0.455\,MHz - 0.455\,MHz + f_{unknown} = 14.755\,MHz - 0.455\,MHz \tag{9.14}$$

$$f_{unknown} = 14.300\,MHz \tag{9.15}$$

We come up with another "possible" culprit – 14.300 MHz. Great algebraic problem solving, but that's our desired frequency, so that's definitely not the right answer!

Receiver system		Direct Sampling Superheterodyne
Intermediate frequency		36kHz
Sensitivity	SSB/CW	(BW: 2.4kHz at 10dB S/N) 1.8–29.999MHz 0.16μV (Preamp 1 ON) 50MHz 0.13μV (Preamp 2 ON)
	AM	(BW: 6kHz at 10dB S/N) 0.5–1.8MHz 12.6μV (Preamp 1 ON) 1.8–29.999MHz 2.0μV (Preamp 1 ON) 50MHz 1.0μV (Preamp 2 ON)
	FM	(BW: 15kHz at 12dB SINAD) 28.0–29.7MHz 0.5μV (Preamp 1 ON) 50MHz 0.25μV (Preamp 2 ON)
Squelch sensitivity (Threshold)	SSB	Less than 5.6μV
	FM	Less than 0.3μV (HF: Preamp 1 ON, 50MHz: Preamp 2 ON)
Selectivity (sharp filter shape)	SSB	(BW: 2.4kHz) More than 2.4kHz / −6dB Less than 3.4kHz / −40dB
	CW	(BW: 500Hz) More than 500Hz / −6dB Less than 700Hz / −40dB
	RTTY	(BW: 500Hz) More than 500Hz / −6dB Less than 800Hz / −40dB
	AM	(BW: 6kHz) More than 6.0kHz / −6dB Less than 10.0kHz / −40dB
	FM	(BW: 15kHz) More than 12.0kHz / −6dB Less than 22kHz / −40dB
Spurious and image rejection ratio	HF	More than 70dB 50MHz: More than 70dB (Except for ADC Aliasing)
Audio output power		More than 2.5W (at 10% distortion with an 8Ω load, 1kHz)

Figure 9.1: Receiver Specifications for ICOM IC-7300. ©2024 Icom America Inc.

Chapter 10

Interference Calculations

The calculations in this lesson are similar to those you did for the image response question, but now we're dealing with *intermodulation interference*, so we're going to end up with two frequencies that might be causing the interference.

While image response is a product of the IF section, intermodulation can happen anywhere in the transmitter-to-receiver system that is *nonlinear*. In practice, there's no such thing as a perfectly linear circuit, so intermodulation interference (aka intermodulation distortion) can rear its ugly head anywhere.

The most likely locations are areas where there are several strong signals. Commercial transmitter sites are at the top of the list, since they are usually hosting multiple stations. Those sites are not where they are by accident, they are in locations that provide maximum coverage the area they serve. That means they're prime sites for ham repeaters. Ideally, one would do these intermodulation calculations *before* renting expensive tower space!

What transmitter frequencies would cause an intermodulation-product signal in a receiver tuned to 146.70 MHz when a nearby station transmits on 146.52 MHz?

This question requires some careful reading. Here's what it's asking. You have a receiver tuned to 146.70 MHz. You're not transmitting, just receiving. Whenever a nearby station transmits on 146.52 MHz, it messes up your reception on 146.70. You hear a garbled up signal that sounds like a combination of two stations. Because you are a smart ham, you recognize this as intermodulation distortion. What are the possible frequencies for that *other* station that's garbling the one at 146.70?

Every transmitter produces spurious emissions – harmonics of the main signal. Ideally, these are sufficiently suppressed that they come out of the antenna at very low – and legal – power, but they're still there. And, lucky you, you just happen to be close enough to that 146.52 transmitter that those harmonics are strong enough for you to pick up. Maybe the intermodulation is coming from that harmonic plus some unknown station. This problem and the formulas should be very familiar from the chapter on algebra:

$$f_{intermodulation} = 2f_1 - f_2 \qquad (10.1)$$

and ...

$$f_{intermodulation} = 2f_2 - f_1 \qquad (10.2)$$

"f_{imd}" means "the frequency where the intermodulation is appearing." – the frequency your receiver is set to, in this case 146.70. f_1 is the frequency of the transmitting station, 146.52, and f_2 is the frequency of the unknown station.

With a little algebraic rearrangement, we get:

$$f_2 = 2f_1 - f_{intermodulation} \qquad (10.3)$$

or ...

$$f_2 = 2(146.52) - 146.70 = 146.34 \qquad (10.4)$$

The second harmonic of 146.52 is $2 \times 146.52 = 293.04$. Well, one hopes your radio isn't picking up a signal at 293.04 MHz when it's tuned to 146.70, so it isn't that – besides, it's intermodulated, so this must be the result of an interaction between that harmonic at 293.04 and another signal. When we combine frequencies, we always end up with a harmonic that is the sum of the frequencies and one that is the difference. (Actually, we end up with more than those two, but in the practical world, we are just concerned with those.) $293.04 - 146.70 = 146.34$. So **146.34** is one possibility.

The other possibility is:

$$f_{intermodulation} = 2f_2 - f_1 \tag{10.5}$$

Why is that "the other possibility?"

We rub a little algebra on that formula. We'll do it a little differently this time, just to show you there are *lots* of correct ways to solve these. First, we'll add f_1 to both sides to get this:

$$f_{intermodulation} + f_1 = 2f_2 \tag{10.6}$$

Then we'll divide both sides by 2

$$\frac{f_{intermodulation} + f_1}{2} = \frac{2f_2}{2} = f_2 \tag{10.7}$$

...or...

$$f_2 = \frac{146.70\,MHz + 146.52\,MHz}{2} = 146.61\,\text{MHz} \tag{10.8}$$

The correct answer for this question is **146.34 MHz and 146.61 MHz**. Beware the answer 146.88 MHz. Yes, you can come up with that number by doubling your receiver frequency then subtracting 146.52, but that makes no sense – your receiver isn't creating any harmonics that would come close to matching the strength of a received signal.

Now, let's simplify things, so you don't have to repeat all that algebra to solve these. Here are the formulas transformed:

$$f_2 = 2f_1 - f_{intermodulation} \tag{10.9}$$

and...

$$f_2 = \frac{f_{intermodulation} + f_1}{2} \tag{10.10}$$

$f_{intermodulation}$ is the frequency to which the transmitter is tuned, f_1 is the "nearby station."

Noise Floor

Another key performance factor for a receiver is the receiver's *noise floor*.

The noise floor is a measure of the noise density (dBm/Hz) or the noise power in a receiver's output with a bandwidth of 1 Hz. It's the sum of all the components that aren't part of the useful signal. Put, we hope, more simply, it's all the output of a receiver within a specified bandwidth that is not the desired signal. Typically, receivers are rated with "noise floor at 1 Hz."

The wider the bandwidth we listen to, the more noise we'll hear. Think of noise as water behind a dam. If you have a tiny crack in the dam, only a little of the water gets through. Turn that crack into something 10 times as wide, you'll get 10 times more water getting through.

How much does increasing a receiver's bandwidth from 50 Hz to 1,000 Hz increase the receiver's noise floor?

For the moment, don't be concerned with what a "noise floor" might be; all this question is asking is "how many dB equal a 20 times gain?" ($1000\,Hz \div 50\,Hz = 20\times$).

Mathematically, what we'll do is reverse what we did for the last question. That means finding the logarithm of the Power Factor, 20; ($\log_{10}(20)$.) We'll multiply that figure by 10 to convert from bells to decibels; remember, a decibel is 1/10th of a bell.

$$dB\ change = \log_{10}(Power\ Factor) \times 10 = \log_{10}(20) \times 10 = 13\,dB \quad (10.11)$$

Chapter 11

Resonant Frequency of a Circuit

We wouldn't have much in the way of electronics without resonant circuits. We use them to create or to emphasize signals we want, and we use them to block signals we don't want. They're in our transmitters, our antenna tuners, and our receivers. Even our antennas are a form of resonant circuit.

While there are many forms of resonant circuits, we'll look at very basic series and parallel resonant circuits that contain an inductor, a capacitor, and a resistor. These are often known as LCR circuits.

Inductors easily pass low frequencies and oppose high frequencies. Capacitors oppose low frequencies and easily pass high frequencies. Either type of component's opposition is called its reactance and, clearly, the reactance is dependent on the frequency being applied to the component. Reactance is measured in ohms and can be substituted for resistance in Ohm's Law calculations.

Resonance in a series resonant circuit occurs at a frequency where the reactance values of the inductor (L) and the capacitor (C) combine to give the lowest impedance possible for the circuit. Put another way, for more advanced students, the resonant frequency of a series resonant circuit is the frequency at which the phase shifts between voltage and current created by the reactances cancel out and voltage and current are in phase.

Given the complexity of what's going on in a resonant circuit, the formula for calculating the resonant frequency of a resonant circuit is really pretty simple!

$$f = \frac{1}{2\pi\sqrt{L \times C}} \tag{11.1}$$

That formula works for both series and parallel resonant circuits, even though the effects of each are almost the precise opposite of the other.

A *ideal* series resonant circuit *passes* its resonant frequency and blocks all other frequencies. In reality, series resonant circuits pass a range of frequencies around the resonant frequency and reduce all others.

An ideal parallel resonant circuit *blocks* its resonant frequency and passes all others. Of course, in reality, parallel resonant circuits *reduce* a range of frequencies around their resonant frequency and pass all others. At resonance, a parallel resonant circuits voltage and current are shifted +90∘ by the inductance and -90∘ by the capacitance, effectively canceling each other.

Having the L and C in that square root bracket (a *radical*) means that changes in those values will have less effect on the final value than if they were just plain old L and C. In other words, you can change L and/or C a lot but f won't change much. If L and C were squared, that would mean even a small change in either value would change f a lot.

At resonance, the reactances of the L and C are equal.

Any equation can be graphed on *rectangular coordinates*. A rectangular coordinates graph consists of an x axis, that runs horizontally through the center of the graph, and a y axis that runs vertically through the center of the graph. Some equations, such as "$x = 31$" or "$x + 4 = 7$", graph as a single point, because there's only one solution. Equations with more unknowns graph as various curves, showing all the possible solutions, or at least enough of the solutions to imply the rest.

Figure 11.1's graphs show what happens to x as a plain value, as a square root, and as a square. To plot this, we'd pick a number on the x axis, then perform some operation on x and plot the result above the x axis. You can see the values of $\sqrt{x}$ increase much more slowly than the other two.

When we graph reactance against frequency, inductive reactance ($2\pi fL$) forms a straight line but capacitive reactance ($\frac{1}{2\pi fC}$) forms a curved line. You can see an example in Figure 11.2. The straight, solid

(a) Values of $y = x$ (b) Values of $y = \sqrt{x}$ (c) Values of $y = x^2$

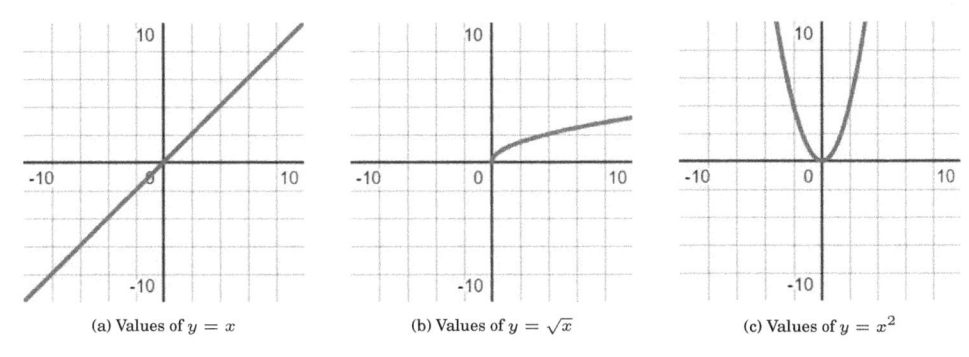

Figure 11.1: Values of $y = x$, $y = \sqrt{x}$, and $y = x^2$

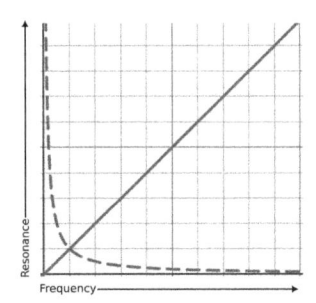

Figure 11.2: Reactance Values

line is inductive reactance and the curved, dotted line is capacitive reactance. The point where those two lines meet is the resonance point of the circuit.

To find the resonant frequency in Hz, also known as simply "the resonance," we multiply the inductance in Henries times the capacitance in Farads, take the square root of that, multiply that square root by 2π (roughly 6.28) then divide that answer into one. Let's try one.

Application

What is the resonant frequency of a series resonant circuit where R is 22 ohms, L is 50 microhenries, and C is 40 picofarads?

We plug in our values. Notice the "R" value isn't used in these calculations, we're just going to work with L and C for the resonant frequency.

Before we get going on this, a reminder; pico = 10^{-12}, micro = 10^{-6}, and milli = 10^{-3}.

$$f = \frac{1}{2\pi\sqrt{L \times C}} = \frac{1}{2\pi\sqrt{(50 \times 10^{-6}\,H) \times (40 \times 10^{-12}\,F)}} \approx 3.56\,MHz \qquad (11.2)$$

The keystrokes are shown in Figure 11.3.

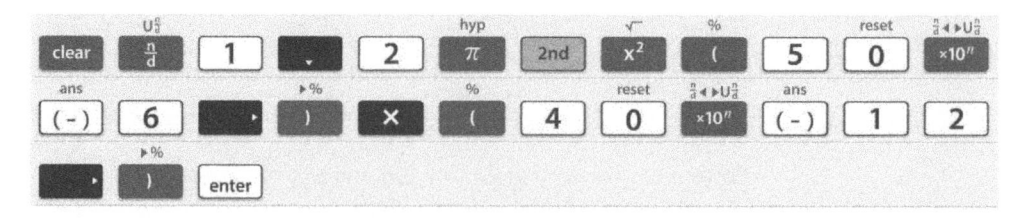

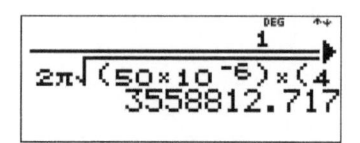

Figure 11.3: Resonance Calculation

We know it looks like a lot of keystrokes, but if you're working through these with calculator in hand, which is what you're doing if you're serious about this, we think you'll quickly get the hang of what you're doing and won't need to read the keystrokes, just the equation; that one you wrote down on paper before you started pushing buttons.

Being able to calculate the resonance of a circuit with known L and C values is all well and good, but what if you are designing a circuit and want to know the L and C values for a given frequency?

Let's design a resonant circuit for 1 kHz. That's a commonly used test tone for many applications. We'll need a capacitor and a coil. What values should we use?

There is a huge range of values that, on paper, produce a circuit with 1 kHz resonance. That range narrows as we take into account the practicality of the components.

In ancient times, before computers, electrical engineers would consult cleverly constructed charts known as nomographs that would quickly give them the values they needed for a particular resonance or any number of other difficult-to-calculate values. You can see one at: www.rfcafe.com/references/.

We can't solve for both variables, L and C, at the same time. With some calculus we could solve for a range of those variables, but that's a lot of unnecessary heavy mathematical lifting. A little semi-educated guesswork plugged into some simple formulas will do the job for us here.

The formula for L, the inductance, when C and f are known is:

$$L = \frac{1}{4\pi^2 f^2 C} \tag{11.3}$$

The formula for C, capacitance, when L and f are known is:

$$C = \frac{1}{4\pi^2 f^2 L} \tag{11.4}$$

From a practical standpoint, it is probably better to choose an off-the-shelf value for the capacitance and a variable inductor for the inductance, since the chances of both precise values we need being readily available from electronic parts suppliers are somewhere between slim and none. With a variable component we can fine-tune that 1 kHz tone to perfection. (Of course, if we used a variable capacitor with a wide range, we could make a tone generator that covered a wide range of frequencies. Decisions, decisions ...) Variable capacitors can have their values affected by all sorts of factors, including humidity, temperature, and even dust. If we want real precision, we'll opt for a variable inductor which is more likely to stay reasonably close to its nominal value. (We could also build our own coil to any value we want, which is a time-honored traditional activity for hams.)

If we just happen to have a 1 pF (picofarad) capacitor in that old coffee can where we throw spare parts, what size inductor would we need to use to make our 1 kHz oscillator?

$$L = \frac{1}{4\pi^2 f^2 C} = \frac{1}{4\pi^2 \times (10^3 Hz)^2 \times 10^{-12} F} = 25,330\,H \tag{11.5}$$

25,330 henries is a lot of henries. What would this 25,330 henry coil look like?

There is no simple formula for coil design. We have in our library a 1943 edition of *Radio Engineers' Handbook* that states, "The design of coils with magnetic cores is normally carried out in a cut-and-try process." Not much has changed since 1943, but we do have online coil calculators now with which we can do our cutting and trying in the digital realm. A bit of fiddling with one (www.allaboutcircuits.com/tools/) revealed a coil about 1 meter long by 1 meter in diameter with 1125 turns of wire around a high-end ferrite core. I was unable to locate a source for such a ferrite core, but if we use an iron core the coil gets even bigger. We'd also need a Freightliner to carry it around. Put simply, this coil is ridiculous.

Let's try plugging in a more reasonable value for C and find a more useful value of L for this 1 kHz tone generator. We'll go for 1 millifarad, $1 \times 10^{-3} F$; a billion (10^9) times more capacitance than a picofarad, which is 10^{-12}.

$$L = \frac{1}{4\pi^2 f^2 C} = \frac{1}{4\pi^2 \times (10^3)^2 \, Hz \times 10^{-3} \, F} = 25.33 \times 10^{-6} H \approx 25 \mu H \quad (11.6)$$

25 microhenries of inductance seems like it should be much more practical to create than the 25,000 henries we needed with that 1 picofarad capacitor, and, indeed, ready-made $25 \, \mu H$ variable inductors are $8.21 from a leading supplier.

Chapter 12

Calculating Reactance

There are at least two more values we need to calculate if we want to do something useful with a resonant circuit. To get to those, we need to know the reactances of the components.

Capacitors and inductors both oppose alternating currents, and this opposition varies with the value of the component and with frequency. We call this opposition Reactance, and in formulas it gets the letter X. It is measured in ohms (Ω). For inductors, as the frequency gets higher, the reactance gets higher, as you see in Figure 12.1a. At the lowest frequency, DC, the inductor looks, essentially, like a straight wire.

For capacitors, as the frequency gets higher, the reactance gets lower, as in Figure 12.1b. At DC, the capacitor looks just like an insulator.

Since these components behave in opposite ways relative to frequency, we need one formula for inductive reactance and another for capacitive reactance. Let's start with inductive reactance, known in formulas as X_L.

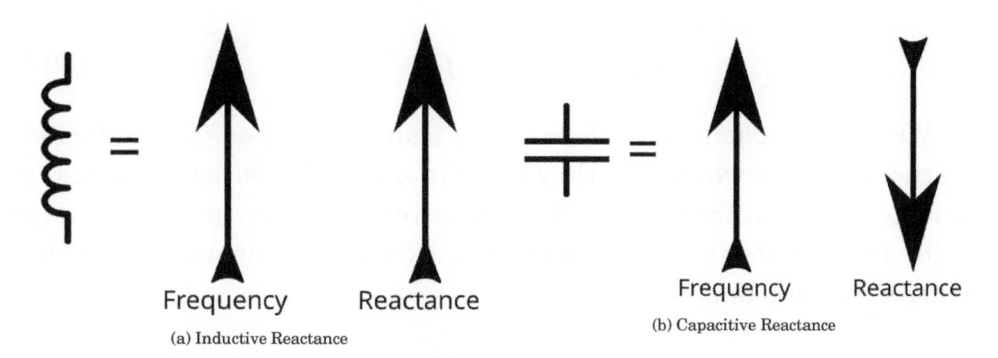

Figure 12.1: Inductive and Capacitive Reactances vs. Frequency

$$X_L = 2\pi f L \qquad (12.1)$$

Inductive reactance equals two times pi times the frequency times the inductance. Notice, if the frequency or the inductance increases, the reactance increases.

> Stop. Appreciate that equation for a moment. It precisely predicts the result of a quite complex interaction of a particular frequency of electromagnetic wave with a coil of a particular value with just four characters. It's like a poem about electricity! It took us 24 words to write a sentence that was a very vague version of the same thing in prose.

If you're curious about the (2π) that keeps showing up in our formulas, that relates to *radians*. Radians are an angular measurement used in place of degrees in a lot of scientific work to vastly simplify formulas like this one. There are, of course, 360 degrees in a circle, and there are 2π radians. One cycle of a sine wave "rotates" through 360 degrees or – yep – 2π radians. A physicist would write the formula differently:

$$X_L = \omega L \qquad (12.2)$$

"ω" (a lower-case Greek "omega") is a physics symbol for "angular speed"; "how quickly is this thing changing its angle of travel?" When we're talking about waves,

$$\omega = 2\pi f \qquad (12.3)$$

We know capacitors work in a way that's the opposite of inductors. Mathematicians would say they work in a way that's the reciprocal of inductors, and so the formula for capacitive reactance is the reciprocal of the formula for inductive reactance. (Reciprocal means "divide that whole thing into one.")

$$X_C = \frac{1}{2\pi f C} \qquad (12.4)$$

> What happens to the ohms of reactance if the frequency or the capacitance increase? Take a short visualization break. Try adding some animation to your visualization to show the relationship between X_C and $2\pi fC$.

Capacitive reactance equals two times pi times the frequency times the capacitance in Farads, all divided into one.

Notice in this formula as the frequency or capacitance increases, the reactance decreases. Bigger numbers on the bottom make the fraction smaller; 3/4 is greater than 3/8. Keeping that in mind seems to help me keep lots of these formulas straight in my head – of course, that requires understanding the physics behind the formula, such as "inductors pass low frequencies and block high frequencies." Once you're clear on that, though, then it makes sense that X_L would increase as either the frequency or the inductance increase.

What's useful in these – and, really, any other formula – is to get all the elements into standard units, such as farads, hertz, henries, and ohms. Let's do one. Let's calculate the reactance of an 82 millihenry coil at 2.65 MHz.

The formula for inductive reactance is

$$X_L = 2\pi f L \qquad (12.5)$$

We'll plug in our values:

$$X_L = 2\pi \times 2,650,000\, Hz \times (82 \times 10^{-3}\, H) \qquad (12.6)$$

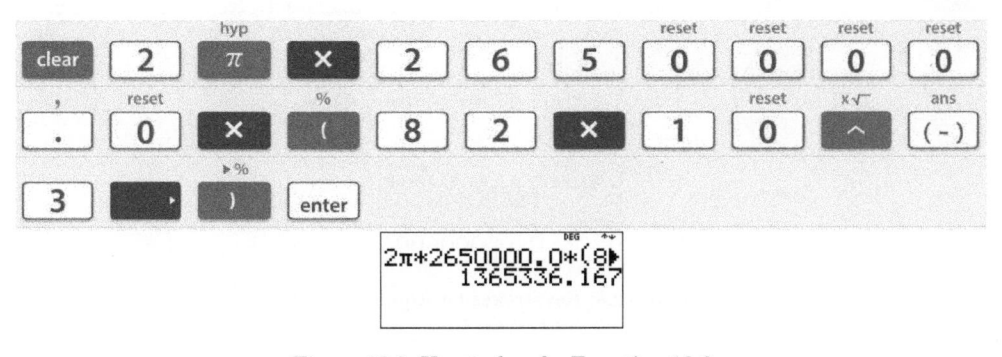

Figure 12.2: Keystrokes for Equation 12.6

(You need to enter 2650000.0 or the calculator will give you the answer "434600π." Another accurate but useless answer.)

You can also use that [×10ⁿ] key to cut down all those digits in the frequency, if you choose to do so.

$$X_L = 2\pi \times (2.65 \times 10^6 \, Hz) \times (82 \times 10^{-3} \, H) \tag{12.7}$$

Are you looking at 1365336.167? Excellent! That's ohms, because we used hertz and henries in our calculations. We'll call it 1.4 megohms.

Let's figure the reactance for a 10 millifarad capacitor at 2.65 MHz. The formula is:

$$X_C = \frac{1}{2\pi f C} \tag{12.8}$$

Plug in the values:

$$X_C = \frac{1}{2\pi(2,650,000 \, Hz) \times (10 \times 10^{-3} \, F)} = 6 \times 10^{-6} \, \Omega \tag{12.9}$$

While we're here, let us show you another handy TI-30XS button you can use for any formula that's a reciprocal. Remember, a number raised to the power of -1 is the reciprocal of that number. There's a key marked [x⁻¹] just above the "7" key on your keyboard. It takes whatever is stored as the last answer and turns it into the reciprocal of that answer. In other words, it divides the answer into 1. So you could do this:

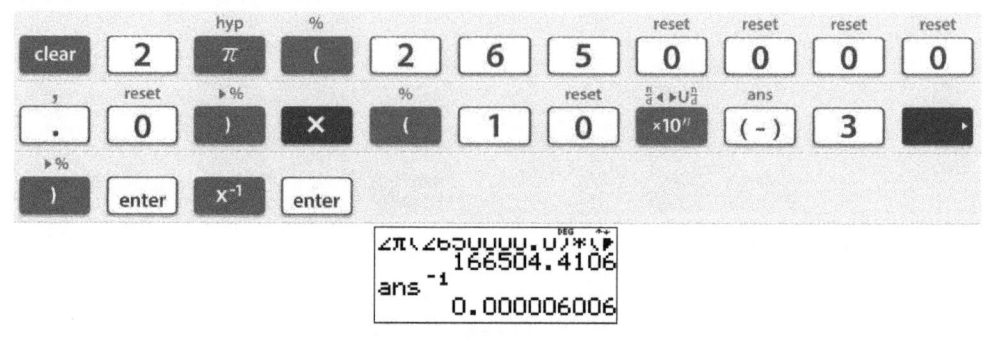

Figure 12.3: Keystrokes for Equation 12.9

It doesn't really save many keystrokes, but some people like the slightly neater readout on the screen.

If you enter the 2.65 MHz as "2650000" and don't end it with a ".0" as you enter the keystrokes, you'll get a strange answer of "53000π." Don't sweat it, just hit your Answer Toggle key ⟨⟩ and you should be looking at 0.000006006, or 6×10⁻⁶ ohms of reactance. For this capacitor, there's only 6 microhms of reactance at that frequency – for all practical purposes, it's a straight wire. Now you know why we have picofarad capacitors.

By the way, the formula for calculating the resonant frequency of a circuit, $\frac{1}{2\pi\sqrt{L \times C}}$ comes directly from these reactance formulas. It is an algebraic rewriting of $2\pi fL = \frac{1}{2\pi fC}$. In other words, resonance occurs when the reactances of the inductance (X_L) and the capacitance (X_C) are equal.

Application

What is the reactance (X_L) of an 40 mH inductor at 7.25 MHz?

The inductive reactance formula is quite straightforward; multiply 2, π, the frequency and the inductance together. Our biggest challenge is getting all the units right. The common wavelength formula, $\frac{300}{f_{MHz}}$, specifies the frequency must be in MHz, but that's an exception. Generally speaking, you're safe converting everything to SI units.

$$X_L = 2\pi fL = 2\pi \times (7.25 \times 10^6 \, Hz) \times (40 \times 10^{-3} \, H) = 1.8 \times 10^6 \, \Omega \quad (12.10)$$

This inductor's reactance is 1.8 megohms at 7.25 MHz. If you were looking for an inductor that would present a very high impedance to a 40-meter signal, this one would work!

What is the reactance (X_C) of a 450 microfarad capacitor at 7.25 MHz?

$$X_C = \frac{1}{2\pi fC} = \frac{1}{2\pi \times (7.25 \times 10^6 \, Hz) \times (450 \times 10^{-6} \, F)} = 48.8 \times 10^{-6} \, \Omega \quad (12.11)$$

This capacitor presents a 48.8 microhm impedance at 7.25 MHz. That's a tiny amount of reactance. For all practical purposes, this capacitor presents zero impedance at that frequency. If we wanted a circuit to resonate at 7.25 MHz, this capacitor could be a good choice for the circuit but the inductor in the problem above would be a terrible choice. To make a resonant circuit with this capacitor you'd need an L of about 1 microhenry (μH).

Chapter 13

The Q Factor

The Q factor of a resonant circuit is a way to express the bandwidth of the frequencies affected by the circuit. If you're familiar with high-fidelity terminology, you could think of it as the measure of the circuit's frequency response. The lower the Q, the wider the bandwidth affected. If we graphed the frequencies passed by a relatively high Q series resonant circuit, it might look something like Figure 13.1a,

A relatively low Q series resonant circuit's frequencies passed would look more like Figure 13.1b. (Consider both of those illustrations to be sketches – the actual curves are more complex.)

For parallel resonant circuits, those curves would be upside down relative to the series curves.

The calculation for the Q of a resonant circuit is where we use that R number we left out of the resonant frequency calculation. Well, you knew it was going to show up somewhere, right?

For series resonant circuits, the formula for Q is:

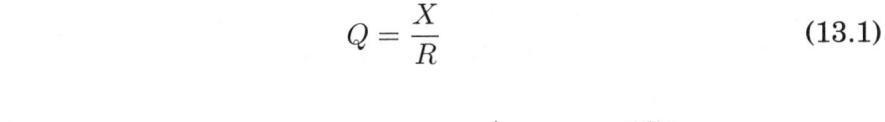

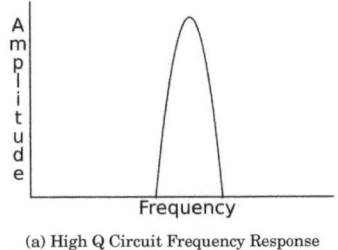

$$Q = \frac{X}{R} \qquad (13.1)$$

(a) High Q Circuit Frequency Response

(b) Low Q Circuit Frequency Response

Figure 13.1: Effects of Q on Frequency Response

Q equals the reactance *at the resonant frequency* divided by the resistance in series in the circuit. As reactance increases, the Q increases, and as resistance increases, the Q decreases.

We know in a resonant circuit we have X_L and X_C – which X do we use? Either one! At the resonant frequency, X_L and X_C are equal.

So let's say we have a reactance of 100 ohms and a resistance of 50 ohms. That gives us a Q of

$$Q = \frac{X}{R} = \frac{100}{50} = 2 \tag{13.2}$$

You might ask, "Two whats? Two ohms, two volts, two Q's.....???" None of those. The Q factor is what they call "a dimensionless term." It's just a factor, a useful number to apply to other calculations, really.

How did we get that formula? The "behind-the-scenes" formula for Q is $Q = \frac{Power_{stored}}{Power_{dissipated}}$. Here's the transformation:

$$Q = \frac{Power_{stored}}{Power_{dissipated}} = \frac{I^2 X}{I^2 R} = \frac{X}{R} \tag{13.3}$$

As generations of electronics students know, "Twinkle, twinkle little star, power equals I squared R." We can substitute X for R in the upper half of the formula. Then the I^2's cancel and we're left with $\frac{X}{R}$.

Parallel resonant circuits are a bit more complicated. The formula above, $Q = \frac{X}{R}$, can be applied to parallel resonant circuits in which the resistor is in series with the inductor. If the resistance is in parallel with the inductor and the resistance has a "very high value," the formula flips upside down.

$$Q = \frac{R}{X} \tag{13.4}$$

"Very high value" is undefined in any reference we could find. If in doubt, you can fall back on another formula;

$$Q = 2\pi \frac{L}{R_s} \tag{13.5}$$

Where R_s is the total resistance in the system.

Application

What are the resonant frequency, the reactance at resonance, and the Q of the series resonant circuit shown in Figure 13.2 (page 75) if L_1 is 50 mH

$$L_1 \quad C_1 \quad R_1$$

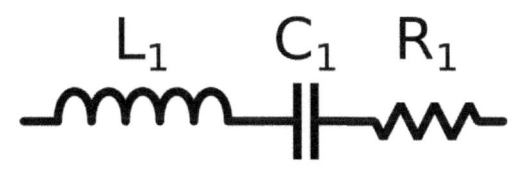

Figure 13.2

(millihenries), C1 is 500 pF (picofarads), and R1 is 10 ohms?
To begin, we'll need the resonant frequency of this circuit.

$$f_{resonant} = \frac{1}{2\pi\sqrt{L \times C}} = \frac{1}{2\pi\sqrt{(50 \times 10^{-3}\,H) \times (500 \times 10^{-12}\,F)}} = 31{,}831\,Hz \quad (13.6)$$

Now that we know the resonant frequency we can calculate the reactance at resonance. Remember, the inductive and capacitive reactances will be equal at resonance.

$$X = 2\pi f L = 2\pi \times 31{,}831\,Hz \times (50 \times 10^{-3}\,H) = 10{,}000\,\Omega \quad (13.7)$$

This circuit has 10,000 ohms of reactance at the resonant frequency of 31,831 Hz.

With the reactance calculated we can move on to the Q of this circuit.

$$Q = \frac{X}{R} = \frac{10{,}000}{10} = 1000 \quad (13.8)$$

We're still missing some important information about this filter. What is its bandwidth? Since it is a series resonant circuit, it will pass some band of frequencies and block the rest; how wide is that band of frequencies that will be passed? Specifically, how wide is the band of frequencies that are 3 dB down from the highest value? We can divide the resonant frequency by the (very high) Q to determine the half-power bandwidth of this circuit.

$$BW_{half-power} = \frac{f_{resonant}}{Q} = \frac{31{,}831\,Hz}{1{,}000} = 31.83\,Hz \quad (13.9)$$

> That's a new formula – you know what to do.

At last that Q number tells us something useful!
The half-power bandwidth of this filter is an extremely narrow 31.83 Hz! If that is too narrow for our purposes, we could bump the resistor's

(a) Resonant Circuit Frequency Response

(b) Half-Power Bandwidth

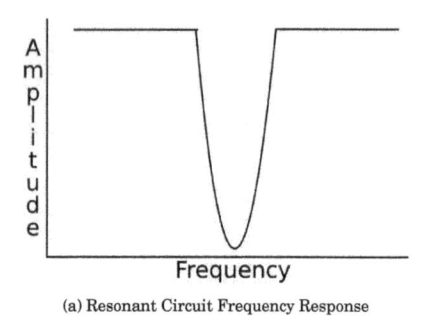

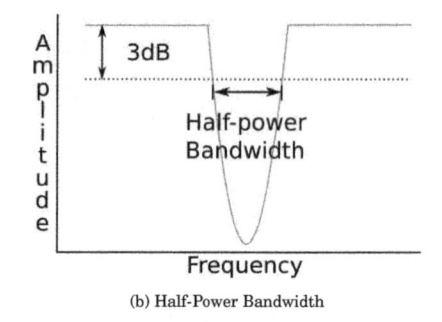

Figure 13.3: Parallel Resonant Circuit Frequency Response & Half-Power Bandwidth

value up to 100 Ω for a half-power bandwidth of 318.31 Hz. For a pretty decent SSB filter, we could make R1 1,000 Ω and have a half-power bandwidth of 3183.1 Hz. You can see that the controlling factor of Q is the resistance in the circuit.

This also points up why we'll never be able to build an analog filter with a zero Hz half-power bandwidth. Notice how the bandwidth decreases with each decrease in resistance. To build a zero Hz half-power bandwidth filter, which would filter one and only one frequency, we'd need a zero resistance circuit. Aside from very exotic laboratory settings, that's not possible. There will always be stray, "parasitic", resistance in our components, in the wiring (or printed circuit board traces) between the components, and even in the connectors that provide the input and output of the circuit.

What is the half-power bandwidth of a parallel resonant circuit that has a resonant frequency of 7.1 MHz and a Q of 150?

This question asks about a parallel resonant circuit. We know a parallel resonant circuit opposes the flow of the resonant frequency. Let's say we have a high Q parallel resonant circuit and its frequency response looks like Figure 13.3a.

The half-power bandwidth is the frequency difference between the lowest frequency that is dropped to half power, which is − 3 DB, and the highest frequency that is dropped to half power, as shown in Figure 13.3b.

Here's how to calculate that bandwidth. Happily, it's a simple formula, one you already saw on page 75.

$$half\ power\ bandwidth = \frac{f_{resonant}}{Q} \qquad (13.10)$$

The half-power bandwidth equals the resonant frequency divided by the Q. As the Q increases, the half power bandwidth gets narrower.

Our first step is to convert 7.1 MHz to Hz. We do that by moving the decimal point 6 places to the right. Why? Because $1\ MHz = 1 \times 10^6 Hz$. We're just multiplying 7.1 by 1,000,000. We could also leave it as 7.1 MHz and *divide* the Q number, 150, by 1,000,000 to get 0.00015. That also leads to a correct answer.

$$Bandwidth_{half\ power} = \frac{7,100,000 Hz}{150} = 47,333\ Hz \approx \mathbf{47.3\ kHz} \quad (13.11)$$

The half-power bandwidth of this filter circuit is 47.3 kHz.

Another factor that comes into play with high Q values is ringing. Ringing is usually an unwanted effect, since it is another form of distortion. (It is used to advantage in certain types of power supplies and amplifiers.) Ringing is the oscillations that continue in the circuit after the initial impulse energizes the circuit. The higher the Q, the more ringing will be present. The reason is that the resistance in the circuit serves to damp those extra vibrations, causing them to lose energy faster than they would if there was no resistance in the circuit.

Think of the resistance in the circuit like the shock absorbers in your car's suspension. Most suspensions consist of springs and shock absorbers. The springs let the suspension flex to smooth out the ride and keep the tires on the road. A shock absorber's job is to damp that suspension system. Without shock absorbers your car would be continually bouncing up and down; the automotive version of ringing. (In fact, a way to test your shock absorbers is to push down sharply on your car's bumper then release it just as sharply. The bumper should come back up to its original height very little bouncing. If it goes "boing, boing, boing" a few times, you need new shocks.)

To avoid ringing altogether, ζ (*zeta*), the *damping ratio*, must be equal to or greater than 1. Damping ratio is one of the more complex pieces of mathematics we will cover in this book. To make matters even more interesting, there are at least two ways of calculating damping ratio (known as "damping factor" in some texts) that yield slightly different answers!

If we know the values of all of the components in the circuit, we can use the following formula:

$$\zeta = \frac{R_{total}}{2\sqrt{\frac{L}{C}}} \quad (13.12)$$

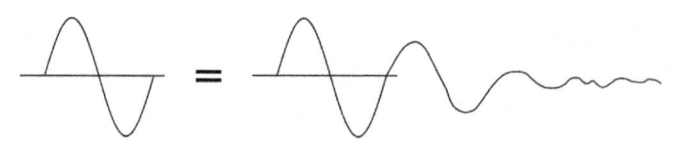

Figure 13.4: Input Signal (L) vs. Signal Produced by Ringing (R)

Since we know the values of the first example in this section, let's plug in those values.

$$\zeta = \frac{R_{total}}{2\sqrt{\frac{L}{C}}} = \frac{10\,\Omega}{2\sqrt{\frac{50\times 10^{-3}\,H}{500\times 10^{-12}\,F}}} = 0.0005 \qquad (13.13)$$

Our resonant circuit has a damping ratio of 0.0005. What does this mean? Figure 13.4 shows the difference between the input to a resonant circuit and the output that results from ringing in the circuit. Ideally, of course, there would be zero ringing and the waveform of the output would match the waveform of the input. A damping ratio of 1 would produce zero ringing.

The damping ratio tells us by how much each sine wave after the first sine wave will diminish in amplitude relative to the preceding sine wave. In other words, in this resonant circuit, each cycle of ringing will be equal to the amplitude of the previous cycle minus that amplitude times 0.0005. If the initial cycle has an amplitude of 1 mW, the next will have an amplitude of 0.9995 mW. It will take thousands of cycles of ringing for the amplitude of this ringing to even get down to 3 dB lower than the original!

If this resonant circuit is meant to be an oscillator, we are not terribly concerned about this. That oscillator is just supposed to keep putting out a single frequency forever. Ringing? Who cares?

If it is a resonant circuit in a Class C amplifier, we're quite pleased; those boxes use a resonant circuit to "fill in" the parts of the waveform that the transistor does not amplify. A circuit with zero ringing would not work in this machine. (This one's ringing might be a bit excessive for a Class C amplifier, but you get the idea.)

On the other hand, if this is a filter for a signal that will be changing frequencies constantly, such as a voice signal, we'd most likely be deeply concerned. If the signal is a phase shift modulated digital signal, this is a complete mess – the receiver will be hearing multiple phase shifts simultaneously and that means the transmission will be, at best, riddled with errors.

You can see that ringing in a resonant circuit is not a trivial issue.

That would seem to raise the question, "What can we do about the damping ratio?" The answer is we can lower the value of L, raise the value of C, and/or raise the value of R. If we lower the value of L, we must also raise the value of C in order to keep the same resonant frequency. L and C have a proportional relationship; if we halve the value of L, we need to double the value of C. Let's see how that works out with the circuit that had an L of 50 mH, a C of 500 pF, and and R of 10 Ω which produced a damping ratio of 0.0005.

We started with these values:

$$\zeta = \frac{R_{total}}{2\sqrt{\frac{L}{C}}} = \frac{10\,\Omega}{2\sqrt{\frac{50\times 10^{-3}\,H}{500\times 10^{-12}\,F}}} = 0.0005 \quad (13.14)$$

We'll halve the value of L and double the value of C. Let's be sure that still produces the same resonant frequency of 31,831 Hz.

$$f_{resonant} = \frac{1}{2\pi\sqrt{L\times C}} = \frac{1}{2\pi(25\times 10^{-3}\,H)\times(1000\times 10^{-12}\,F)} = 31,831\,Hz \quad (13.15)$$

That checks out. Now, what about that damping ratio?

$$\zeta = \frac{R_{total}}{2\sqrt{\frac{L}{C}}} = \frac{10\,\Omega}{2\sqrt{\frac{25\times 10^{-3}\,H}{1000\times 10^{-12}\,F}}} = 0.001 \quad (13.16)$$

(We realize that the value 1000×10^{-12} should more properly be written as 1×10^{-9}. We have kept that value at picohenries for clarity.)

By changing the L and C values we have doubled the damping ratio to 0.001.

Let's say we're still not satisfied. Perhaps we're building an audio filter and we'd like it to have a reasonably high damping ratio of 0.2. Dividing 0.2 (the desired ratio) by 0.001 (the ratio produced by the values above) tells us we need to lower the inductance by a factor of 200 and raise the capacitance by the same factor. Let's see what we get.

$(25 \times 10^{-3}) \div 200 = 125 \times 10^{-6}$ and $(1000 \times 10^{-12}) \times 200 = 200 \times 10^{-9}$.

$$\zeta = \frac{R_{total}}{2\sqrt{\frac{L}{C}}} = \frac{10\,\Omega}{2\sqrt{\frac{125\times 10^{-6}\,H}{200\times 10^{-9}\,F}}} = 0.2 \quad (13.17)$$

Mission accomplished! Or is it? What is the Q of this circuit? The X in $\frac{X}{R}$ has plummeted.

$$X_L = 2\pi f L = 2\pi \times 31831\,Hz \times (125 \times 10^{-6}\,H) = 25\,\Omega \quad (13.18)$$

Now we can calculate the Q:

$$Q = \frac{X}{R} = \frac{25\,\Omega}{10\,\Omega} = 2.5 \tag{13.19}$$

We could also have used the original L and C values of 50 mH and 500 pF and multiplied the R value by 400 to get the same damping ratio of 0.2 with the same Q of 2.5. (Not a coincidence.) That wildly lower Q will rather dramatically affect the bandwidth of the circuit.

With a Q of 2.5, what is the half-power bandwidth of this circuit?

Remember, this was a circuit with a Q of 1,000! What has happened to our half-power bandwidth?

$$BW_{half-power} = \frac{f_{resonant}}{Q} = \frac{31{,}831\,Hz}{2.5} = 12{,}732\,Hz \tag{13.20}$$

Now this circuit's half-power bandwidth has gone from about 32 Hz to 12,732 Hz. If what we wanted was a very (!) wideband resonant circuit, we're brilliant designers. If not – it's back to the drawing board. At least we have computers and calculators for this work. Imagine the days when most of this was accomplished with a slide rule.

Resonant circuits are systems, each part linked to every other part. Change one value and everything else changes. In fact, the mathematics of resonant circuits goes on and on; it is a very deep rabbit hole. Consider; we have not even touched on such factors as insertion loss or transient response, to name just a couple. This is why we end up with exotic filter designs like elliptical filters, Chebyshev filters, and many more. Some of those designs required the invention of entire new branches of mathematics. However, in the spirit of our often repeated proverb, "In ham radio, optimum is optional," we will end this exploration here and leave the student to go spelunking in that world on their own.

Chapter 14

Time Constants

Resonant circuits work because different value components charge and discharge at different rates. The term for that pace is *time constant*, and it is represented by the Greek letter *tau*; ($\tau\tau$).

One time constant for capacitors is defined as the time required, in seconds, for the capacitor in an RC circuit to be charged to 63.2% of the applied voltage, or to discharge to 36.8% of its initial voltage.

The formula for calculating the time constant of a capacitor is:

$$\tau_C = RC \qquad (14.1)$$

The τ is equal to the resistance in the circuit, in ohms, multiplied by the capacitance in farads. The bigger the resistance or the capacitance, the bigger the τ, and the longer it will take to charge the capacitor.

Inductors have time constants, too. Here's the formula:

$$\tau_L = \frac{L}{R} \qquad (14.2)$$

In this formula, increasing the inductance increases the time constant, but increasing the resistance decreases the time constant.

Recall that in the first instant of current flow through an inductor, the inductor opposes the flow of current. As the magnetic field around the inductor stabilizes, more current flows.

Notice, capacitor time constants tell us how long it takes to reach 63.2% of the maximum *voltage* across the capacitor. Inductor time constants tell us how long it takes to reach 63.2% of maximum *current* flow through the inductor as the inductor.

You're unlikely to be doing time constant calculations to figure out resonant circuits. There are far simpler equations to accomplish those

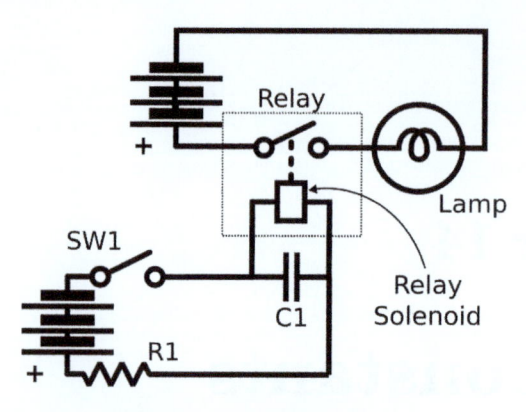

Figure 14.1: Time Delay Circuit

calculations. A slightly more likely application would be something like a timing circuit, something like Figure 14.1.

We'll imagine that the battery on the lower left side of that circuit is a 9-volt battery, the resistor is a 1000-ohm resistor, the capacitor is a 1-millifarad capacitor and that the relay activates and closes the switch it controls when the voltage across it reaches 6.32 volts.[1]

The formula $\tau_C = LC$ tells us the time constant of that capacitor is one second. That means one second after we close SW1 the voltage across the capacitor will be 6.32 volts. Until then, the voltage will be below that value because the capacitor, in effect, acts like a conductor that "steals" the relay's voltage. At 6.32 volts the relay closes, lighting the lamp. We've created a circuit that lights a light one second after we close the switch.

These days, we'd probably build this circuit with a digital timer, but for many, many years this was essentially how time-delay circuits worked.

Let's say the resistance of the lamp is 1000 ohms and we insert a 1-millihenry coil in series in the lamp circuit. Now we have an inductor with a 1-second time constant ($\tau_L = \frac{L}{R}$). Now when we switch on SW1, after one second the lamp will begin to fade up to full brightness over the course of one second.[2]

---

[1] All these values are chosen for convenience in calculation and do not represent a practical circuit. Close, but not quite.

[2] The principal is accurate, but so far as we know that's not how anyone ever built circuits to fade lights up or down.

Application

What is the time constant of a circuit having two 220 microfarad capacitors and two 1 megohm resistors, all in parallel?

Now we get to use our simple formula for the time constant, $\tau = RC$, after we do some slightly more complicated maths to figure out the R and the C.

The question describes this circuit:

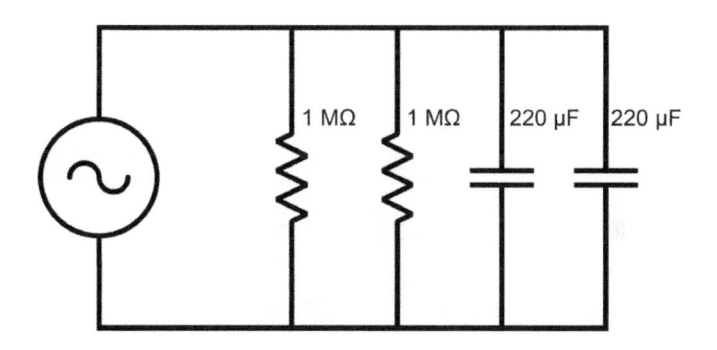

We have two capacitors and two resistors, all in parallel. We need to know the total resistance in the circuit and the total capacitance. There's no special formula for "parallel time constant." It's the same formula as for a series time constant – the total capacitance multiplied by the total resistance. (Sweet!)

The resistors are 1 megohm each, and they're in parallel. Because we wisely and diligently studied parallel circuits, we remember that two equal resistors in parallel add up to one-half the value of one of the resistors, so we have 500 kΩ of resistance, or 500,000 Ω, or 500×10^3.

Capacitors in parallel add together, just like resistors in series, so we have

$$2 \times 220 \ \mu F = 440 \ \mu F \tag{14.3}$$

440 microfarads (μF) of capacitance, also known as 440×10^{-6} or 0.00044 farads.

We've simplified our circuit to this:

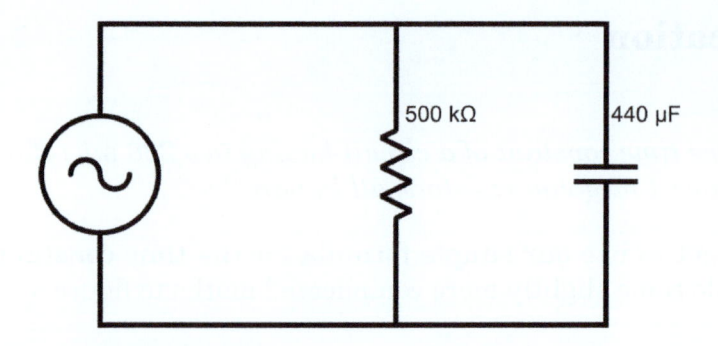

Now we multiply the capacitance by the resistance to find our τ, the time constant in seconds.

$$\tau = RC = (500 \times 10^3) \times (440 \times 10^{-6}) = \mathbf{220 \text{ s}} \qquad (14.4)$$

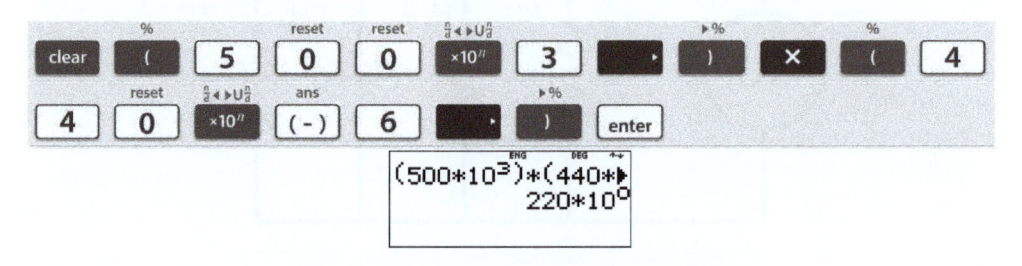

Figure 14.2: Keystrokes for Equation 14.4

We'd use exactly the same formula if the circuit was this:

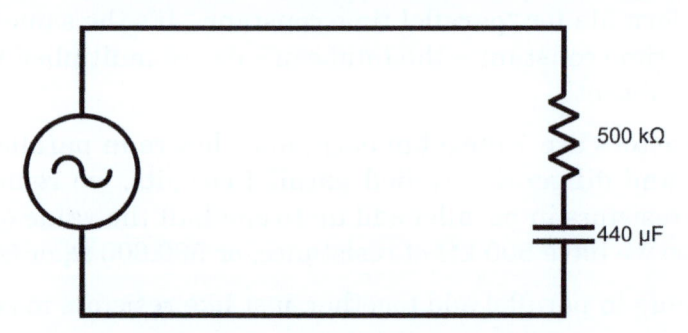

What is the time constant of a circuit having one 5 mH inductor in series with a 20 Ω resistor?

For the time constant of an inductor, we use the formula $\tau_L = \frac{L}{R}$ and the answer we get is how long it takes the inductor to *discharge* 63.2%.

$$\tau_L = \frac{L}{R} = \frac{5 \times 10^{-3} \text{ H}}{20 \text{ }\Omega} = 250 \times 10^{-6} \text{ s} \qquad (14.5)$$

This inductor will discharge to 63.2% of its original charge in 250 microseconds.

Why, you might ask, 63.2%? What does that number have to do with anything? Did they just throw random numbers in a hat and pick one? Nope. Let's take a look at how a capacitor charges. No matter what the value, the voltage across the capacitor is going to create a curve something like this as the capacitor charges.

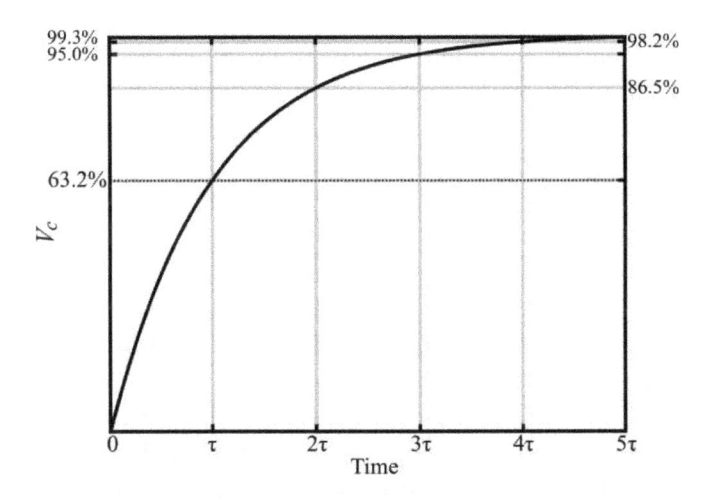

At first, lots of electrons are rushing in, and the voltage is increasing rapidly. Then, as the capacitor fills, it gets harder to push in more electrons – think of blowing up a balloon.

So, let's imagine we're going to charge a capacitor to 100 volts, and let's say the τ of the circuit is 1 second. After the first τ – one second – the voltage across the capacitor is 63.2 volts. After another second, it has only gone up 23.3 volts to 86.5 volts. We wait another second – now the voltage has only gone up 8.5 volts to 95 volts. Another whole second goes by with a gain of only 3.2 volts to 98.2 volts and yet another full second of charging only gets us to 99.3 volts. If it seems like we're never going to quite get to 100 volts at this rate, you are correct!

It's one of those Zeno's Paradox deals. Zeno of Elea was a Greek philosopher who, around 450 BC, reasoned that one could never walk across a room because before one could get across the room, one had to travel half-way across the room, and before one could travel half-way ... well, you can see where this is going. Or perhaps I should say, where it isn't going! Mathematicians call this sort of thing a non-summable infinite series. As it turns out, these turn up all over the physical world, not just in electronics, and the maths to describe these phenomena is based on what are called *natural logarithms*. (It took the invention of calculus in the mid-1700's to finally resolve Zeno's paradox.)

Natural logarithms are based on a mathematical constant known as *e*, or Euler's Constant, which is approximately 2.178. It's sort of a strange number that shows up in electronics, finance, physics, geology, and even biology. And $1 - e^{-1} \approx 63.2\%$. So, aside from its mathematical and engineering significance, it turns out that using that 63.2% point on the curve to define the time constant, τ, spares us from using the natural logarithm (l_n) and e^x keys on our calculator and simplifies our formulas down to $\tau_C = RC$ and $\tau_L = \frac{L}{R}$.

As you ponder these formulas you might wonder, "How did time get in there?" In other words, how can we multiply or divide two seemingly static values and have the answer come out in units of time? After all, there's no amount of mathematical gymnastics that will allow you to multiply the sides of a 2 unit by 4 unit rectangle and somehow get an answer of 8 seconds.

Time snuck in there on the coattails of the resistance. Recall that resistance is equal to $\frac{V}{I}$. The definition of an ampere of current, the I, includes time; an ampere is a quantity of 1 coulomb of charges moving past a point in *one second*. (A coulomb is 6.241509×10^{18} charge units – for simplicity's sake, call it 6.241509×10^{18} electrons. It is an oddball number because it was originally defined in terms of the attraction between two energized wires.) Volts and ohms are quantities,[3] but amperes are a rate and rates are "so many somethings *per* some unit."

---

[3]Mathematicians also call them "scalars" in some contexts.

Chapter 15

Phase Angles

As we advance in electronics, it seems *phase angle* becomes a more and more important concept.

Let's start with the nature of a sine wave. Remember, a sine wave is really just a graph of a circle – you could say it's a graph of a spinning circle. Imagine a wheel that is rotating clockwise once per minute. We'll set that wheel next to a sheet of graph paper, so the center of the wheel is lined up with the x axis.

Let's pick a point on the circle at what would be 9 o'clock on a clock face. We'll take a ruler and at the "zero seconds" point on the x axis, we make a dot. We wait five seconds. Now the point on the wheel has rotated up to the 10 o'clock position. We find the five seconds mark on the x axis, and the point on the y axis that the point is lined up with now. We make a mark above the x axis, to the right of the first mark. Every five seconds, we plot the position of that point on the wheel on our graph. After the wheel rotates 360 degrees, we'll have a plot of a sine curve.

When we talk about a phase angle, we're talking about comparing one position on that circle to another, in terms of degrees of rotation.

Figure 15.1 shows what a 90-degree phase angle between two sine

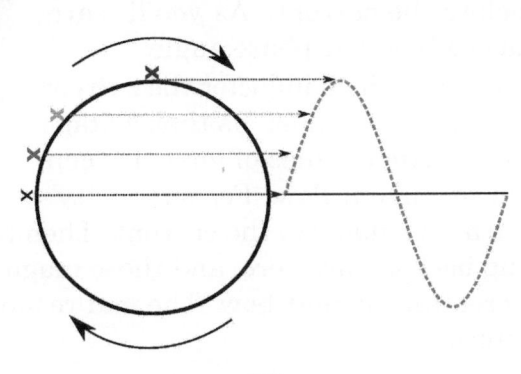

87

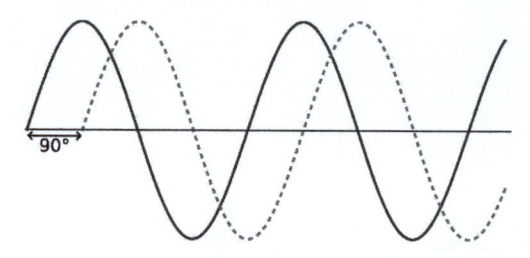

Figure 15.1: Sine Waves 90° Out of Phase

waves (of equal frequency) looks like.

Just as the first sine wave – represented by the solid line – reaches the 90 degree mark, the second sine wave is at the 0 mark and starting up. We call that a "90 degree phase angle." If the sine waves matched up at all points, we'd say they're "in phase", but these are "out of phase", and they're out of phase by 90 degrees.

Now things get a little strange. When we learn DC electronics, we learn that voltage and current always go together – put a higher voltage on a circuit, the current instantly gets higher, because of Ohm's Law. I call it the "garden hose" model of electricity, and you've heard lots of garden hose related similes as you've learned DC electricity. The garden hose model works great. For DC.

AC circuits make us set aside that thinking. In AC circuits, current and voltage can get out of phase. We can have spots where there is current with no voltage, or voltage with no current. Current and voltage can be put out of phase by capacitors and inductors.

So let's say in the illustration above, that the solid line represents voltage and the dotted line represents the current. In that illustration the voltage is *leading* the current by 90 degrees. In other words, the voltage is "happening" before the current. As you'll learn, that means we have a "positive phase angle."

When AC passes through an inductor the current lags behind the voltage; remember, as that voltage is headed up, those magnetic lines of force are generating opposition to current flow. For simplicity's sake, we'll say the inductor looks like an insulator to the current. Then the voltage peaks, and starts heading back toward zero, and those magnetic lines of force start collapsing, creating current flow. The end result is voltage leads current by 90 degrees.

Figure 15.2: Oliver "Happy" Heaviside

This almost seems to contradict Ohm's Law. How can there be current with no voltage? We can return to the garden hose model for this. Imagine the hose is running and you put your thumb over the end. The current (water) will stop while the voltage (pressure) in the hose increases. When you release your thumb, the water flows and the pressure in the hose drops. We could even say that a model of capacitive phase shift would be putting your thumb over the end of an already-flowing hose while the model of inductive phase shift would be putting your thumb over the end just before the water is turned on then releasing it after the pressure builds. This is far from a perfect analogy for what happens in a circuit with reactance, but it does reflect the general idea.

We'll teach you an easy way to remember the relationship of voltage and current through a reactance, but to do so we need to revert to the old symbol for voltage, E.

The summer of 1885 in London, England, was dreadfully hot. Of course, in those days, there were no refrigerators, but there were ice boxes, and there were commercial ice services that would deliver big five pound blocks of ice to homes to keep food fresh.

During that summer, Oliver Heaviside, was sweating away at his desk, coming up with the equations that would lay the foundation for our understanding of reactances and impedances. Heaviside was, by all accounts, absolutely brilliant, but is all but forgotten these days. Among other achievements, he invented coaxial cable, and predicted the existence of the ionosphere. (Another name for the E region is "Kennelly-Heaviside layer.")[1] His Telegrapher's Equations made practical trans-Atlantic telegraphy possible.

Heaviside was a notorious sourpuss and anything but a social butterfly. His social life seems to have consisted entirely of surreptitiously delivering his manuscripts for articles in *The Electrician* magazine. He reportedly never met the editor for whom he worked for years. What seems to have been a natural tendency toward being reclusive was no doubt exacerbated by his hearing loss in his later years.

Every week, Heaviside's meditations would be interrupted by his ice delivery from the Roy G. Biv Ice Company. His ice man was named Eli. Eli's deliveries always irritated Heaviside. One day, though, he suddenly saw that Eli was, unwittingly, the perfect mnemonic for the voltage and current relationships Heaviside was discovering. "Gadzooks!", he cried, "**E**lectromotive force (E, as we say in these days before they came up with

---

[1] In T S Eliot's poem *Old Possum's Book of Practical Cats* which was the basis of the musical *Cats*, cats go "Up, up, up, up to the Heaviside layer" to leave their old life and begin a new one.

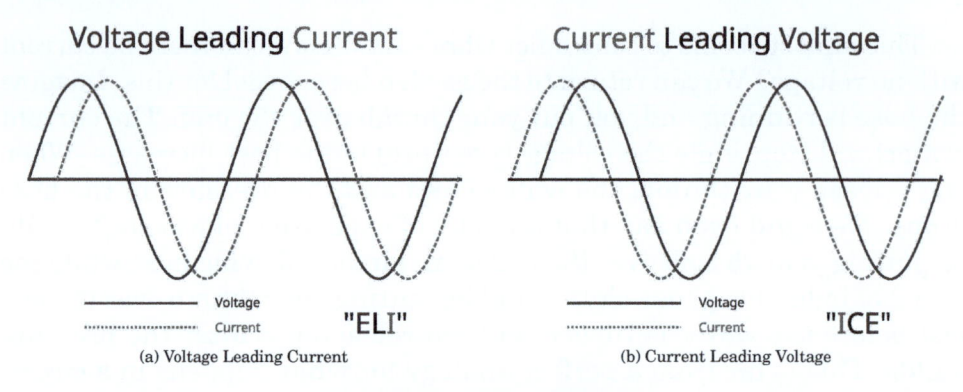

Figure 15.3: ELI the ICE Man

that newfangled "voltage" word) leads current (**I**) across an inductor (**L**) and current (**I**) leads electromotive force **E**) across a capacitor (C) – ELI the ICE man! Oh, jolly good, old bean."

Ever since that fateful (and completely fictitious) day, hams and other electronics students have used "ELI the ICE Man" to remember that across an inductance (across an L) E (voltage) comes before I (current); E leads I. Across a Capacitance, I leads E. Good ol' ELI the ICE Man!

When we say the voltage is *leading* the current, as it does across an inductance, we mean the voltage "happens" before the current. See Figure 15.3a.

We could also describe that state of affairs as "the current is *lagging* the voltage."

When we say the current is leading the voltage, as it does across a capacitance, that means the current happens first, followed by the voltage. (See Figure 15.3b.) Of course, we could also say "the voltage is lagging the current."

Before we get into calculating phase angles and representing impedances graphically, we'll take a quick look at one of the practical applications of this knowledge.

Power Factor

Once a circuit contains reactance, our simple Joule's Law power formula ($Power = Voltage \times Current$) gets complicated by the presence of reactive power.

Put simply, reactive power is the power that is echoing around inside a circuit that contains reactance. You'll recall that capacitors and inductors both store energy; capacitors store it in an electric field, inductors in a magnetic field. Energy cannot be stored and be doing useful work at the same time! You could think of reactive power as being the energy that is stored in the circuit's capacitors and/or inductors.

Reactive power is almost a contradiction in terms. It is "wattless" power. It does no work, it is nonproductive. In an imaginary ideal circuit, it would resonate in the circuit forever, never dissipating.

Figure 15.4 shows two sine waves 90 degrees out of phase relative to each other.

We'll say the solid wave is voltage and the dotted wave is current. Notice how each time the voltage hits maximum positive or negative, the current is at 0. Since power is volts times amps that gives us zero watts at those moments. There are even large parts of the cycle where the current is positive while the voltage is negative, and vice versa.

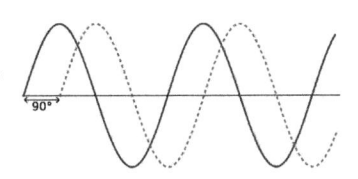

Figure 15.4: Two Sine Waves 90° Out of Phase

In fact, when we look at each point of time on that graph and calculate the real power, it comes out something like the bold line curve on Figure 15.5.

Understand, that bold represents power, and half the time it's actually negative – it's pushing power back to the source! So when we average it all out, we have no real power being used in this circuit at all.

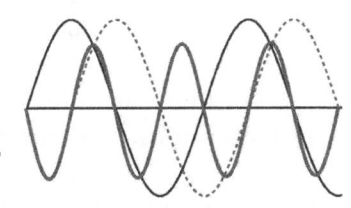

Figure 15.5: Real Power at 90° Phase Difference

At other phase angles, we get more real power and less reactive power. For instance, at 60 degrees of phase angle, it's a 50/50 split between productive power and reactive power.

Phase angles are created by the mix of inductive and capacitive reactances in the circuit.

If we multiply the volts and amps going into the circuit, we get a figure called *apparent power*. But if we factor in how much of that is trapped in reactive power, we get the *real power*. The ratio of real power to apparent power is called the *power factor*, and once we know the phase angle it is easy to calculate that power factor.

$$PowerFactor = \cos(PhaseAngle) \qquad (15.1)$$

The power factor is equal to the cosine of the phase angle. "Cosine" is a trigonometry term, as is "sine", as in "sine wave." The power factor will always be a number between 0 and 1.

If you're using your trusty TI-30XS calculator, you just press the "cos" key, enter the phase angle, add a parenthesis, press enter, and there's the power factor. Unfortunately, if you just enter, say, "30" for the phase angle, the 30-XS will tell you the answer is $\frac{\sqrt{3}}{2}$ which is perfectly accurate and perfectly useless for ham exams, but if you enter the angle as "30.0" it will tell you the power factor for a 30 degree phase angle is 0.866.

One quick calculation to try out is the power factor if the phase angle is 90°. You'll instantly find out the power factor in that case is 0 – NO real power is being consumed in the circuit at all!

We have searched in vain for a ham who has ever, in their ham activities, calculated the *power factor* for a circuit, but it is good to know the principle because it underlies most of our concerns with impedance matching.

That's certainly not to say that there is no practical application of the concept of power factors. Anyone who works in power distribution for an electric company is always concerned about power factor. Electric motors present large inductive loads so electrical substations usually have banks of capacitors to offset those inductive loads and keep voltage and current in phase for the system. You can see one in Figure 15.6.

On a more local level, many of those motors also have built-in "starting capacitors." At start-up, an electric motor presents a very large inductive load, so the capacitor serves to get the voltage and current back in phase to deliver enough power to get the motor started. Most of those motors include a switch that automatically takes the capacitor out of the circuit once the motor is up to speed; keeping it in the circuit would throw the phase angle between voltage and current off in the capacitive direction.

Figure 15.6: Capacitor Bank at Electrical Substation

Application

How many watts are consumed in a circuit having a power factor of 0.6 if the input is 200VAC at 5 amperes?

Ohm's Law tells us that the power – the *apparent* power – of the circuit is 1000 watts (220 $VAC \times 5$ A). Multiply the 1000 watts by the power factor of 0.6 to get the real power figure of **600 watts.**

How many watts are consumed in a circuit in which the reactances create a 45 degree phase angle if the input is 100 VAC at 4 amperes?

This is a three-part calculation. First, we need to calculate the apparent power of this circuit, using Ohm's Law.

$$P = V \times I = 100\ V \times 4\ A = 400\ W \qquad (15.2)$$

Then we calculate the power factor of the circuit. We're given the phase angle, 45°, so we take the cosine of the phase angle to reveal the power factor.

$$Power\ Factor = \cos(Phase\ Angle) = \cos(45) = 0.707 = 70.7\% \qquad (15.3)$$

Now we can calculate the real power consumed in the circuit.

$$P_{Real} = P_{Apparent} \times Power\ Factor = 400\ V \times 0.707 \approx 283\ W \qquad (15.4)$$

A circuit has a power factor of 0.6. What is the circuit's approximate phase angle?

To calculate this we use the reciprocal of the cosine, $\cos^{-1}$, also known as the *secant*. Here's the formula:

$$Phase\ Angle = \cos^{-1}(Power\ Factor) = \cos^{-1}(0.6) \approx 53° \quad (15.5)$$

In the next chapter we will deal with a topic that, based on my conversations with hams over the years, has scared a lot of hams away from the highest level exams. DO NOT PANIC. We will get you through this. You'll see.

Remember, if Michael can do this, you can do this. Michael passed high school algebra mostly on the basis of his good looks and winning personality, such as they are – and probably on the teacher's eagerness to never see him again. So, relax.

Chapter 16

Calculating & Graphing Impedance

As amateur radio operators, our practical concerns about impedance most often revolve around mis-matches somewhere in the transmitter/feed line/antenna system. We'll delve into the mathematics of impedances so you can better understand the process of impedance matching.

Impedance Basics

In this section, you're going to learn how to calculate and graph the impedance and phase angle of a circuit. Let's start with a review of impedance basics.

Impedance is the opposition to current flow in an AC circuit. It is created by the interaction of four values.

FREQUENCY: Resistance, capacitance, and inductance are all fixed values, in the sense that those components are what they are, no matter what circuit they are in. Impedance is not a fixed value; you can't go to the electronics parts store and buy x amount of impedance. Impedance varies with frequency unless the impedance is a pure resistance.[1]

RESISTANCE: A resistor in an AC circuit behaves just the same way a resistor behaves in a DC circuit. It dissipates electrical power as heat. A 50 ohm resistor in an AC circuit is exactly the same as a 50 ohm resistor in a DC circuit.

CAPACITIVE REACTANCE: A capacitor passes high frequencies, and opposes low frequencies.

---

[1] In the real world, there's no such thing as a 100% pure resistance; *parasitic reactance* always sneaks in there somehow. However, at HF frequencies we can usually ignore parasitic reactance.

INDUCTIVE REACTANCE: An inductor passes low frequencies – remember, it's essentially invisible to direct current, which has a frequency of zero – and opposes high frequencies, just the opposite behavior of a capacitor.

It's those characteristics that let us combine inductors and capacitors in various ways to create resonant circuits. Those same characteristics also create impedance mismatches.

Reactance affects the phase relationship of voltage and current.

By controlling the inductance and capacitance in an AC circuit, we control those phase relationships.

When we mathematically combine the Resistance and Reactance in a circuit, we get Impedance, represented in formulas by the letter Z.

Obviously, this is a more complex calculation than $V = I \times R$. It takes a couple of steps to complete. First, we have to calculate the Capacitive Reactance and the Inductive Reactance, then use those to calculate the Impedance.

Resistance is, of course, measured in ohms. So is reactance. So is impedance. While this has been the source of confusion for generations of folks trying to learn electronics, it's critical to our calculations that all the units are the same.

The resistive component of the total impedance is equal to the ohms of resistance, no matter what the frequency. Simple. There is no "resistive reactance", resistors aren't reactive. They're a "pure" impedance. A 50 ohm resistor has 50 ohms of impedance, period. That's not true for inductors and capacitors.

The inductive reactance formula is :

$$X_L = 2\pi f L \qquad (16.1)$$

In the inductive reactance formula, as the frequency or inductance increase, reactance increases. Mathematicians would say that in that formula, frequency, inductance, and reactance have a direct relationship.

Capacitive reactance is the reciprocal of inductive reactance. In a sense, it's the opposite, since capacitors act in a manner that's the opposite of inductors. The formula is:

$$X_C = \frac{1}{2\pi f C} \qquad (16.2)$$

In the capacitive reactance formula, as frequency or capacitance *increase*, the reactance *decreases*. It's what maths folks call an inverse relationship.

To calculate the total reactance of a circuit containing X_L and X_C we subtract the X_C from the X_L. Because of that, some texts teach that the formula for capacitive reactance is what you see in Equation 16.3, where $2\pi fC$ is divided into -1.

$$X_C = \frac{-1}{2\pi fC} \qquad (16.3)$$

That will give you a negative value for X_C.

$$X_{Total} = X_L - X_C \qquad (16.4)$$

It's always "subtract X_C from X_L", never the reverse, because capacitive reactance is a negative value. (If you used the "divide into -1 version of the X_C formula, you would add X_L and X_C.)

On to the last step:

This is the formula for Z, impedance:

$$Z = \sqrt{R^2 + X^2} \qquad (16.5)$$

It is possible to make this a one-step calculation:

$$Z = \sqrt{R^2 + (2\pi fL - \frac{1}{2\pi fC})^2} \qquad (16.6)$$

While accurate, this form of the formula is unnecessarily large and clumsy for practical applications. We show it so you can see how all the pieces go together, but in practice it is much simpler to break the problem into steps as we have shown.

Impedance equals the square root of the resistance squared plus the reactance squared.

With a little of the basic algebra you know, we can rewrite that formula by squaring both sides of $Z = \sqrt{R^2 + X^2}$:

$$Z^2 = R^2 + X^2 \qquad (16.7)$$

Or, since there is no rule that says we can't change the letters, we could say...

$$A^2 + B^2 = C^2 \qquad (16.8)$$

It's the good ol' Pythagorean Theorem! For all the mystery and mathematics that surround the topic, every single impedance problem comes down to nothing but a "what is the length of hypotenuse C" problem, and the length of "hypotenuse C" is equal to the impedance. This means it

(a) Inductive Reactance   (b) Capacitive Reactance

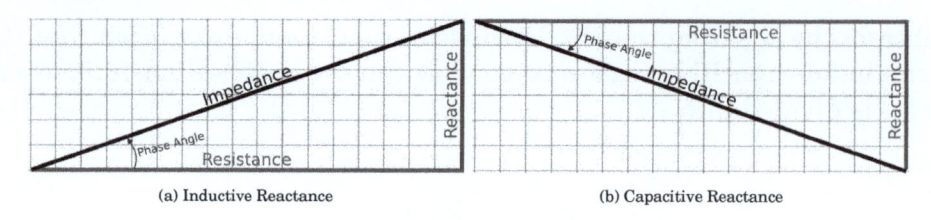

Figure 16.1: Phasor Diagrams

is simple to create a graph of the resistance, reactance, and impedance. See Figure 16.1a

In that illustration we have the resistance plotted as a point on the x-axis of our graph – that's "x" the axis, not "X" for reactance – and reactance plotted on the y-axis. In the example above, the reactance is more inductive than capacitive, so we end up with a positive value for reactance. We also end up with a positive phase angle.

If the reactance was more capacitive than inductive, we'd plot a negative number on the Y axis, and end up with a triangle like the one in Figure 16.1b.

That triangle shows a negative phase angle.

> Spend some time with those triangles. Get clear about the relationships there, especially the positive nature of inductive reactance and the negative nature of capacitive reactance.

Those diagrams have a technical name – they are called "phasor diagrams."

Chances are probably pretty good that you will not often sit down with graph paper and pencil to create a phasor diagram. They are, however, a very clear way to visualize and understand what is happening in any circuit that contains reactances.

Chapter 17

Rectangular & Polar Coordinates

Notice that in both phasor examples in chapter 16, we're really just identifying a point on a graph, which is the point where the impedance and reactance lines meet. We could specify that point in a couple of ways, and it turns out that each different way comes in handy in different situations.

Imagine that point is a destination where you want someone to meet you. You could tell them, "Ralphie, old bean, stand right here at 0,0, what we call the *origin*. Now face due east, so you're looking at all the positive X numbers. You'll walk straight down the X road 400 paces, make a 90 degree turn to the left, then walk up the Y road 300 paces and you'll be there." If you did that, you'd be using what many references call *rectangular coordinates*. Specifically, the 300 and the 400 are the rectangular coordinates.

If Ralphie and you know all about this sort of thing, you don't need all those words – you could shorten the whole thing down to "(400, 300)", and that's how you'd write the instructions for Ralph if you were both mathematicians, engineers, or, say, well-read ham radio operators.

Another way you could get Ralph to the same destination would be to say, "Ralphie, stand right here facing due east. Now turn 36.87 degrees to your left and walk 500 paces and you'll be there." Were you to use that system, Ralph ends up at the same spot, but you have used what are called *polar coordinates* and the you used were 36.87° and 500. If you had told ol' Ralph to turn 36.87° to his right, we'd call that -36.87°.

If you gave Ralph those directions in maths shorthand, you'd just write, "(500, 36.87°)." Ralph would know those were polar coordinates because one of the values is in degrees.

99

In ancient times, polar plots existed only on paper. With the advent of instruments known as Network Vector Analyzers (or, sometimes, Vector Network Analyzers) polar plots have gone electronic, since one way NVA's display their analyses is in polar plots. Antenna azimuth and elevation patterns are polar plots. Later we will learn about Smith Charts, which are another type of polar plot.

If we use polar coordinates, we call that line that forms the hypotenuse – in other words, the impedance line – a "vector." Vectors have a magnitude, also known in electronics as an amplitude, represented by the length, and an angle which is, in this case, the phase angle.

Because exams might ask about both rectangular and polar coordinates, it's important to get the distinction, which is really a pretty simple one; rectangular coordinates specify the points on the X and Y axes, polar coordinates specify the angle and length of the hypotenuse, also known as the vector. Realize, too, that in practice, no matter which set of coordinates we use, we end up with, basically, the same picture and the same values for the reactances, the impedance, and the phase angle. Ralph ends up in the same spot.

In practice, if we start with a known (or desired) impedance and phase angle, we'd use polar coordinates.

When we plot polar coordinates, it's handy to use some special polar coordinate graph paper, so the polar coordinate plot in Figure 17.1 is on some of that. It has coordinates of (6.5, 35°).

From there, we can calculate what we don't know, which is the resistance and reactance. It's just a matter of a tiny bit of trigonometry to calculate the rest of the triangle.

If we know the resistance and reactance, we'd use the rectangular coordinate system as shown in Figure 17.2, and we'll give this one coordinates of (4,3).

From there, we can calculate the length of the hypotenuse, from (0,0) to (4,3), and the phase angle.

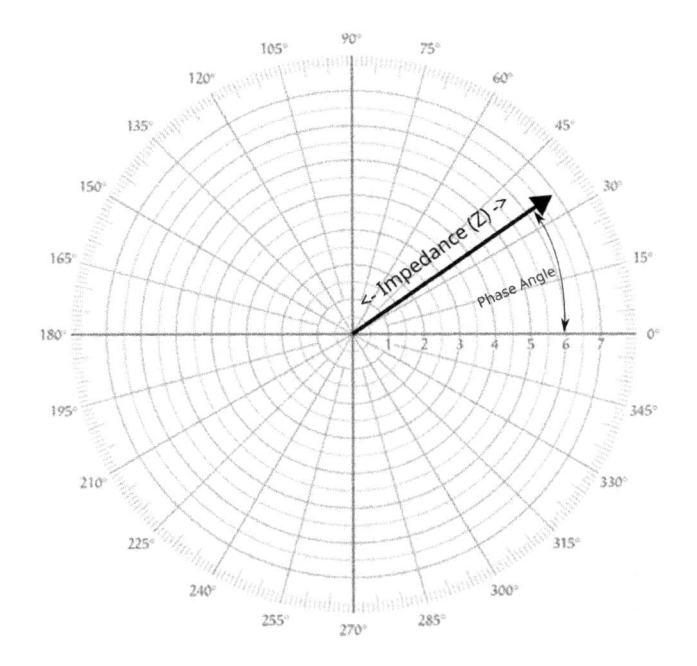

Figure 17.1: Polar Coordinates

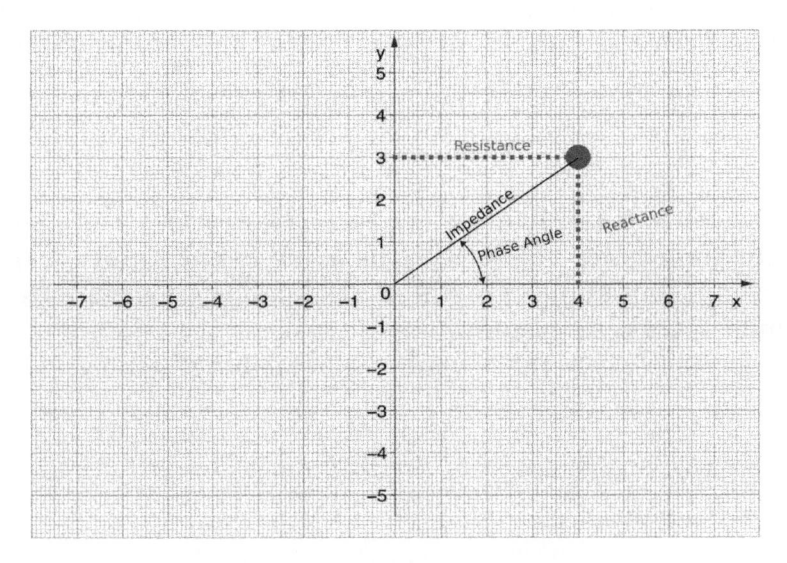

Figure 17.2: Rectangular Coordinates

101

Impedance Matching

In most impedance matching scenarios, we're trying to *eliminate* the reactance then *adjust* the resistance to the impedance value needed. If you imagine either the rectangular coordinates plot or the polar coordinates plot that results from eliminating the reactances, you'll see it reduces the phase angle to zero. Remember, *reactive power* is *non-productive power* so zeroing out that phase angle is a Good Thing.

If our impedance is inductive, we add some capacitance. If the impedance is capacitive, we add some inductance.

At least in theory, if the resistance is too low, we can add some resistance in series. If it is too high, we can add some in parallel. (In practice, we very seldom add resistance to an antenna system – that would be a waste of watts.)

Adding or subtracting reactance from the circuit is most of what's happening inside an "antenna tuner." If you open one, you'll probably find a couple of variable capacitors and an inductor inside. The inductor might or might not be variable, depending on the design. Any other circuitry you find inside is probably for driving the meter on the front.

Application

For application examples, we'll borrow a couple of questions and an illustration from a recent US Amateur Extra Class license exam.

Which point on Figure 17.3 best represents the impedance of a series circuit consisting of a 400-ohm resistor and a 38-picofarad capacitor at 14 MHz?

You don't need to actually draw the triangle, just locate the point that marks the impedance value. It's the same circuit in every question, but the component values are different. All of these questions use the rectangular coordinate system.

Let's do the maths – the formula for inductive reactance is:

$$X_C = \frac{1}{2\pi f C} \qquad (17.1)$$

Plug in the values:

$$X_C = \frac{1}{2\pi \times (14 \times 10^6 \, MHz) \times (38 \times 10^{-12} \, F)} = 299.16 \, \Omega \approx j300 \, \Omega \qquad (17.2)$$

Figure E5-1

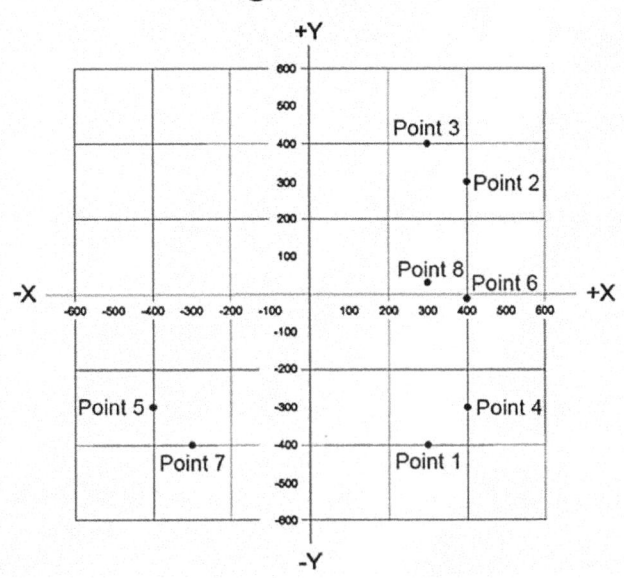

Figure 17.3: US Amateur Extra Class Exam Figure E5-1

That should give you an answer of 299.16 ohms – call it 300 ohms of capacitive reactance, or, more formally, -j300, since capacitive reactance is always considered negative. The total reactance in the circuit, $X_L - X_C$ is (0 − 300 = −300), since there is no X_L in this circuit.

Let's send Ralphie on a walk. Our coordinates for this impedance are (400, -j300) Plot the 400 ohms of resistance on the X axis,

103

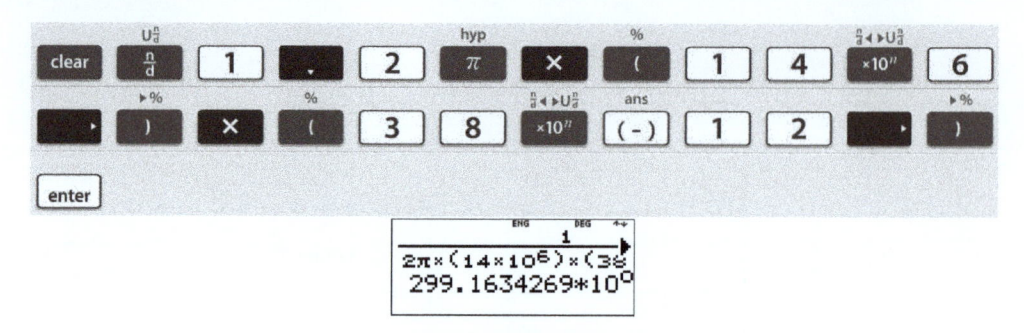

Figure 17.4: Keystrokes for Equation 17.2

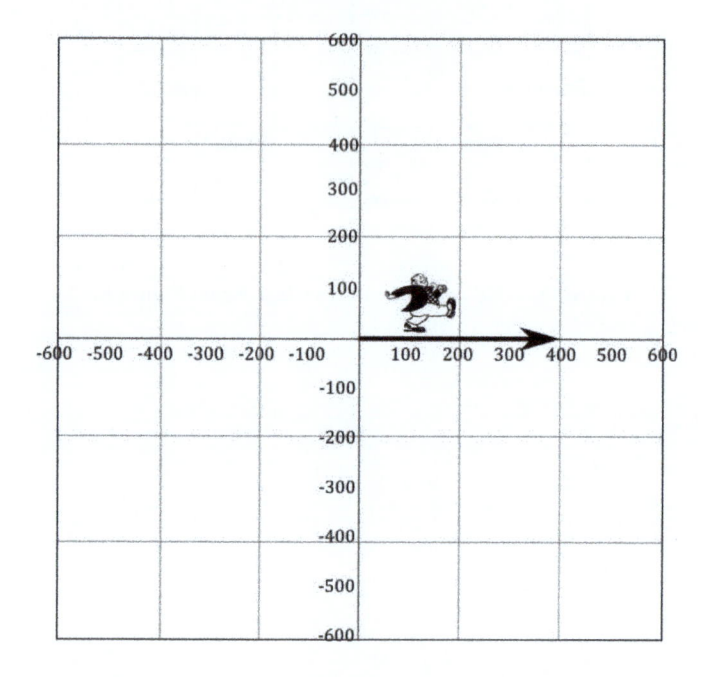

then go straight *down* to the point opposite -300 on the Y axis, where you'll find **point 4**.

Is there a shortcut for this question? There is if you understand the maths of impedance.

Let's work this out.

We'll plot the resistance on the X axis. In this case, we have 400 ohms of resistance, and since resistance is always a positive number, right away we can narrow down the possible correct points on Figure E5-2 to those lined up with the +400 on the X axis, and those are, from top to bottom, points 2, 6, and 4. Wouldn't you know it, they included all of those points in the answers, so there's no shortcut available ...yet.

Ah, but wait! There's no inductor in this circuit, only a capacitor.

104

Capacitive reactance is *always* a negative value. That tells us the correct answer *must* be on the negative side of the Y axis, the part below the X axis. There's only one point at 400 on the X scale with a negative value on the Y scale and it is **point 4**.

Which point in Figure 17.3 best represents the impedance of a series circuit consisting of a 300-ohm resistor and an 18-microhenry inductor at 3.505 MHz?

    A. Point 1
    B. Point 3
    C. Point 7
    D. Point 8

This circuit has 300 ohms of resistance, so go to 300 on the X axis. We have already narrowed down the possible correct answers to three points, points 3, 8 and 1.

There's only inductive reactance in this circuit, because there's only a coil, no capacitor. That tells us the correct answer must be on the positive half of the Y axis – uh, oh, points 3 and 8 are both positive. We're not done. We'll need to calculate the inductive reactance.

The formula for inductive reactance is $X_L = 2\pi f L$. We know the "f", which is 3.505 MHz, or 3,505,000 Hz, and we know the L, 18 microhenries, or 18×10^{-6} henries, or 0.000018 henries.

$$X_L = 2\pi f L = 2\pi \times (3.505 \times 10^6\ MHz) \times (18 \times 10^{-6}\ H) = 396.41\ \Omega \approx 400\ \Omega \quad (17.3)$$

We don't need a lot of precision on this answer – call it 400 ohms.

$$X_L - X_C = 400 - 0 = 400\ \Omega = +j400\ \Omega \quad (17.4)$$

In a very scientific and precise fashion, we put our finger on the 300 on the X axis – for the resistance of 300 ohms – and go straight *up* until we're even with the +400 on the Y axis – for 400 ohms of inductive reactance – and there's **point 3**.

Which point on Figure 17.3 best represents the impedance of a series circuit consisting of a 300-ohm resistor and a 19-picofarad capacitor at 21.200 MHz?

    A. Point 1
    B. Point 3
    C. Point 7
    D. Point 8

In this problem, we have 300 ohms of resistance. That puts us at 300 on the X axis. Once again, we have three possible points.

Let's calculate the capacitive reactance in this circuit.

$$X_C = \frac{1}{2\pi f C} \tag{17.5}$$

$$X_C = \frac{1}{2\pi \times (21.2 \times 10^6 \: Hz) \times (19 \times 10^{-12} \: F)} = 395.1215 \: \Omega \approx 395 \Omega \tag{17.6}$$

We have 395 ohms of capacitive reactance in this circuit, so:

$$X_L - X_C = 0 \: \Omega - 395 \: \Omega = -395 \: \Omega \: or -j395 \: \Omega \tag{17.7}$$

Our coordinates are (300, -j395.)

Can we do this one without calculating the reactance? This circuit has no inductor in it, only a resistor and a capacitor, so we know that our correct answer must be in that lower right-hand quadrant where we have positive values of X and negative values of Y. There's only one point at 300 on the X axis and a negative number on the Y axis, and that's **point 1.**

What is the phase angle between the voltage across and the current through a series RLC circuit if XC is 500 ohms, R is 1 kilohm, and XL is 250 ohms?

  A. 68.2 degrees with the voltage leading the current
  B. 14.0 degrees with the voltage leading the current
  C. 14.0 degrees with the voltage lagging the current
  D. 68.2 degrees with the voltage lagging the current

To calculate the phase angle of a series RLC circuit, we need the capacitive reactance (X_C), the inductive reactance (X_L), and the resistance. They've given us all the values we need in the question, and all we need to do is plug them into the right formula. Here's the right formula:

$$Phase \: Angle = \tan^{-1}(\frac{X_L - X_c}{R}) \tag{17.8}$$

What a great place for a visualization break, right?

"Tan⁻¹" is the reciprocal of "tangent." It's $\frac{1}{Tangent}$. Tangent is a trigonometry function, used to calculate the angle formed between the "adjacent" and "hypotenuse" sides of a right triangle when the "opposite" and "adjacent" sides are known.

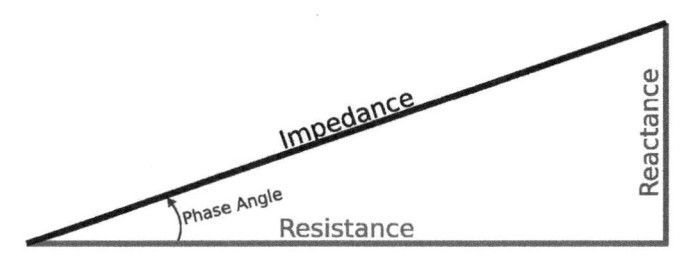

Figure 17.5: Resistance, Reactance, Impedance, and Phase Angle

The answer comes out in degrees. If the degrees are a negative number, that means the current is leading the voltage or, put another way, the voltage is lagging the current. Think of it this way: The sign of the angle always refers to voltage. If the sign is positive, the voltage is *ahead of* the current. It's winning the race. If the sign is negative, the voltage is *behind* the current – it's losing the race.

So let's plug in our numbers. If you're using the TI-30XS, look at the "tan" key. Above it, you'll see "tan-1", and you'll be using that key by pressing the "2nd" key in the upper left hand corner – the bright green one – then the "tan" key.

$$Phase\ Angle = \tan^{-1}(\frac{250\ \Omega - 500\ \Omega}{1000\ \Omega}) = -14.03624347° \approx -14.0° \qquad (17.9)$$

The sign of the phase angle tells us where the voltage is relative to the current. A negative phase angle means the voltage is behind the current.

The answer is the phase angle is **14.0 degrees with the voltage lagging the current**.

We've given you the exact keys we pressed to get this answer in Figure 17.6.

As it does with certain other commands, the TI-30XS automatically fills in the opening parenthesis after "tan^{-1}."

After you press Enter you should be looking at that -14.036 number. That's all there is to it. Congratulations, you were just doing trigonometry.

If you can't seem to find the "tan^{-1}" key on your scientific calculator, look for "arctan", for "arctangent." It's the same function, just a different way to write it.

Now let's step back a moment. Does that negative phase angle make sense?

$$\tan^{-1}\left(\frac{250-500}{1000}\right)$$
$$-14.03624347$$

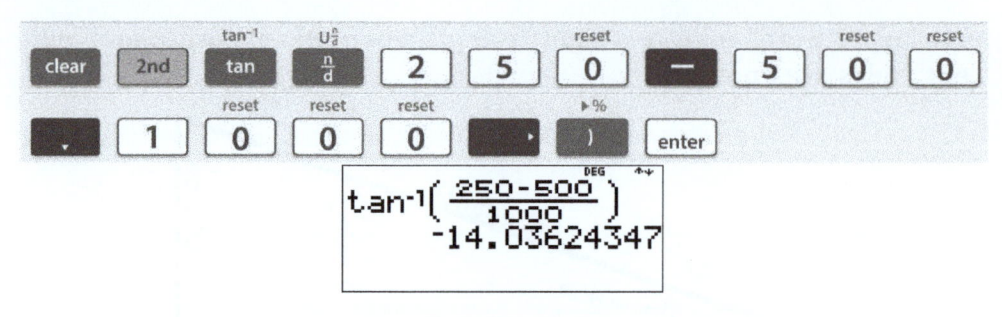

Figure 17.6: Keystrokes for Equation 17.9

We have more capacitive reactance than inductive reactance, so we have a net capacitive reactance. ELI the ICE man tells us that across a capacitance, voltage lags current, so yes, negative 14 degrees makes sense.

Impedance Matching With a Transformer

Our most common method of impedance matching, an antenna tuner, is really a form of transformer. The side connected to the transmitter "looks like" 50 ohms of impedance to the transmitter, and the side connected to the feed line and antenna looks like whatever it needs to look like to those parts of the system.

We can also use an actual transformer to match impedances. This is less common because it is only practical over limited frequency ranges and because it is typically more lossy than the LC circuit common in antenna tuners, but it has its applications.

The question, then, is what impedances should each side of the transformer have in order to achieve a match? Let's say we're trying to match a 25-ohm antenna to a 50-ohm transmitter output. How many turns of wire should be on the primary coil and how many on the secondary?

The answer will be a ratio. On paper, if that turns out to be a 5:1 ratio, it doesn't matter if the primary has five turns and the secondary has one turn or if it is 500 turns on the primary and 100 on the secondary. (In practice, that's very definitely not the case, but we're talking principles here.)

If we're using a transformer to transform a voltage, the ratio is simple. A 2:1 primary to secondary turns ratio will yield a 2:1 ratio of voltages. Put 100 volts into the primary, 50 volts comes out the secondary.

When impedance matching, it is not that simple. Just about any time we're talking impedance, squares and square roots are going to be involved, and, sure enough, the solution for impedance matching transformers involves a square root. Specifically, it's:

$$\frac{N_{Primary}}{N_{Secondary}} = \sqrt{\frac{Z_{Source}}{Z_{Load}}} = \sqrt{\frac{Turns_{Primary}}{Turns_{Secondary}}} \quad (17.10)$$

N = the Number of turns of the primary and secondary transformer coils. The Z's, Z_{Source} and Z_{Load}, are the impedances we want at the secondary and primary. Notice this formula is "upside down" relative to the formula we used for the output voltage of a transformer, where the secondary winding was on the top of the fraction.

Application

What is the turns ratio of a transformer used to match a vacuum tube RF amplifier having 600-ohm output impedance to a feed line having 50-ohm impedance?

We're going to hook the output of some *source* – in this case an RF amplifier – to the primary winding of a transformer. We're going to hook the output of the secondary winding of that transformer to something else – in this question, a feed line – which we'll call the *load*. The question wants to know what the ratio of turns between the primary and secondary windings should be to achieve an impedance match between the source and the load.

$$\frac{N_{Primary}}{N_{Secondary}} = \sqrt{\frac{Z_{Source}}{Z_{Load}}} = \sqrt{\frac{Turns_{Primary}}{Turns_{Secondary}}} \quad (17.11)$$

$$\frac{N_{Primary}}{N_{Secondary}} = \sqrt{\frac{Z_{Source}}{Z_{Load}}} = \sqrt{\frac{600\Omega}{50\Omega}} = \sqrt{12} = 3.46 \quad (17.12)$$

109

$$\sqrt{\frac{600.0}{50}}$$
$$3.464101615$$

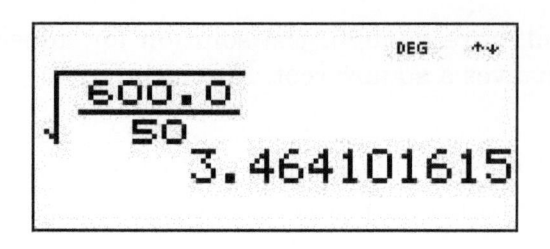

If you wanted to go step-by-step, here's the procedure. Zp = 600. Zs = 50. 600/50 is 12. The square root of 12 is **3.46**. The ratio of Np to Ns will be **3.46 to 1**. We need 3.46 turns on the primary winding for each turn on the secondary winding.

Chapter 18

Op-Amp Circuits

Op-amps are *operational amplifiers*. *Operational* because their original purpose was to perform arithmetic operations in computers. They're no longer widely used in that application, but designers have found hundreds of other uses for this versatile component. It is safe to say that you would be hard-pressed to purchase an electronic product that does not contain at least one op-amp.

A more accurate description of an op-amp is *inverting differential amplifier.* *Inverting* because in most applications the amplifier reverses ("inverts") the polarity of the signal. *Differential* because op-amps amplify the difference between the signals applied to the inputs. Less difference reduces the gain of the amplifier. One input is called the inverting input. The other is the non-inverting input.

Op-amp gain is set by two resistors, one ahead of the non-inverting input – the "-" input in Figure 18.1 – and the other looping back from the output to the non-inverting input. The first resistor we will call R_1, the second is R_F, the F being for "feedback." Since the op-amp is an inverting amplifier, that means the signal coming through R_F is out of phase relative to the input, so it cancels some part of the input. The stronger the signal from R_F, the more the gain is *reduced*, since higher values of the resistor reduce the feedback signal's strength. The ratio between those two resistors, $\frac{R_F}{R_1}$, is the gain of the op-amp.

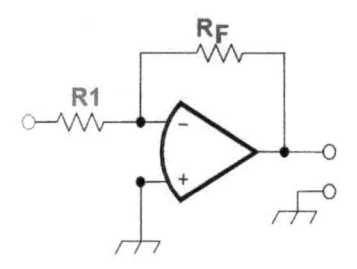

Figure 18.1: Op-Amp

Application

What voltage gain can be expected from the circuit in Figure 18.1 when R1 is 10 ohms and RF is 470 ohms?

We set the gain for an op-amp by changing the values of the two resistors you see in the schematic above, R_1 and R_F. The ratio between the values of those two resistors will equal the gain.

$$Gain = \frac{R_F}{R_1} \tag{18.1}$$

You can keep this formula straight in your mind by thinking about the function of R_1 and remembering that an op-amp is a differential amplifier – it's going to amplify the difference between the non-inverting input and the inverting input. The higher the value of R_1, the less difference there is between those inputs, so R1 reduces gain. The higher the value of R_1 in the formula, the lower the gain. R_F is a negative feedback source. If a negative voltage goes into the input, then R_F is feeding a positive voltage to the input, so the lower the value of R_F, the lower the gain.

For the question, then, it's

$$Gain = \frac{R_F}{R_1} = \frac{470}{10} = 47 \tag{18.2}$$

Chapter 19

Modulation Index & Deviation Ratio

In this section, we explore the not-so-deep philosophical question, "What does 100% modulation mean on FM?"

We will cover calculations for *modulation index* and *deviation ratio*. That makes it seem like there are two formulas to learn here, but, in fact, modulation index and deviation ratio are just different terms for essentially the same thing so there's just one formula. Different exams and references use different terms, but they are interchangeable in almost all instances.

Formally, there is a tiny difference in the formulas.

$$Modulation\ Index = \frac{Frequency\ Deviation}{Modulation\ Frequency} = \frac{\Delta f}{f_{mod}} \quad (19.1)$$

$$Deviation\ Ratio = \frac{Max\ Frequency\ Deviation}{Max\ Modulation\ Frequency} = \frac{\Delta f_{max}}{Max\ F_{mod}} \quad (19.2)$$

You could calculate the modulation index for any frequency you might transmit. In practice, we very seldom care about anything except the ratio produced by that maximum modulation frequency because that leads us to the maximum bandwidth of the signal.

An *index* is simply a percentage divided by 100. So an index of 1.0 = 100%. An index of 0.8 = 80%, and an index of 250 = 250%.

With AM, it's easy to see the 100% modulation point with an oscilloscope. Since an unmodulated carrier just looks like a sine wave, we're at 100% when the modulation envelope scrunches that sine wave down to zero.

But what about FM? How much frequency change is 100%? 100% of what? 1 Hz of change? 1kHz? 1 MHz? How about a signal that deviates from, say, the 10 meter band to the gigahertz band? There's no physical law that we couldn't use that kind of deviation. (There's a whole lot of communications law about that, though, not to mention that it's a wildly impractical scheme.) What's 100%? With FM – and other *angle modulated* signals, we need another measure, one that relates to the maximum frequency we want to transmit, since that frequency will give us the widest deviation.

When we speak of deviation, we're talking about how much we change the frequency of the carrier – and that's going to automatically end up being a "plus and minus" number, since the carrier will swing both higher and lower than the center frequency. However, when we do deviation calculations, we don't worry about the "plus and minus" stuff, we're only concerned with how far from center we're going to deviate. (Maths folks would say we want the "absolute value" of the deviation.) So, an "8 kHz deviation" means we're going up 8 kHz and down 8 kHz and end up occupying about 16 kHz of bandwidth. Obviously, the formula for bandwidth is *Deviation* × 2.

The formula for modulation index is:

$$ModulationIndex = \frac{\Delta f}{f_{\max}} \qquad (19.3)$$

In that formula, $f_{\max}$ is the highest frequency component in the transmitted signal (in Hz), and Δf is the carrier's maximum deviation (also in Hz) from its center frequency. (The Greek letter delta, Δ, is commonly used to designate "amount of change.") So if we want to transmit everything up to 8 kHz, and that causes our carrier to change frequency by 8 kHz higher and 8 kHz lower, that's a modulation index of 1. (8000/8000 = 1.)

Interesting to know, but trivial, right? Yes. Until we set a limit on the modulation index. Most (maybe all) countries limit our modulation index to 1.0 on frequencies below 29.0 MHz. Then, we can work it this way:

$$\Delta f = f_{max} \times 1.0 \qquad (19.4)$$

...and now we know by how many Hz our carrier can deviate and, because the index is 1.0, that tells the maximum frequency (f_{max}) we can transmit. Much more useful, though the $f_{(max)}$ number is disappointingly low; for a 3 kHz bandwidth, the maximum frequency we can transmit is 1.5 kHz.

All that still leaves us with a question; why in the world would we ever need a modulation index greater than 1? If we can send, say, a 1 kHz tone with 1 kHz of deviation (occupying 2 kHz of bandwidth) why would we use more deviation and, thus, more bandwidth? The answer is noise. In the next chapter we'll cover the Shannon-Hartley Theorem that deals with this. That theorem tells us that signals with wider bandwidth overcome higher levels of noise than signals with narrower bandwidth. That's why commercial FM broadcasters occupy bandwidths that, to hams, seem extravagant; $\approx$ 200 kHz!

Those wide bandwidths used by commercial broadcasters also make stereo FM possible. Just like AM, FM creates sidebands. Unlike AM, those sidebands are not necessarily carrying the modulation information. FM sidebands are called subcarriers, because they can carry different information than the main channel. Subcarriers can be amplitude modulated, which saves bandwidth. We seldom if ever use subcarriers in ham radio, but commercial broadcasting makes extensive use of them.

Commercial stereo FM sends a monaural signal (Left + Right, or $L + R$) on the main channel, then creates a subcarrier at 38 kHz away from the main channel center frequency. That subcarrier is amplitude modulated with an "L minus R" signal. In the receiver, the $L+R$ and $L-R$ signals are combined to create an L signal and the $L - R$ is subtracted from the $L + R$ to create the R signal. If that looks like algebra to you, you're right; it is even called "algebraic summing."

Since the authorities frown on stations occupying more bandwidth than they are supposed to, adding that stereo signal to a commercial FM signal requires turning down the overall modulation of that signal by about 10 dB to accommodate the extra bandwidth of the L-R signal.

Application

What is the modulation index of an FM-phone signal having a maximum frequency deviation of 3000 Hz either side of the carrier frequency when the modulating frequency is 1000 Hz?

The modulation index we choose to use is more or less arbitrary – it's limited by law below 29.0 MHz to 1.0, but not in the higher frequencies. For this question, they ask about an FM-phone signal having a maximum frequency deviation of 3000 Hz either side of the carrier frequency when the modulating frequency is 1000 Hz.

$$Modulation\ Index = \frac{\Delta f}{f_{max}} = \frac{3,000\ Hz}{1,000\ Hz} = 3 \qquad (19.5)$$

What is the deviation ratio of an **FM** phone signal having a maximum frequency swing of plus or minus 5 kHz if the highest modulation frequency is 3 kHz?

The maths for *deviation ratio* is exactly the same as the maths for *modulation index*. We use the same formula and plug in different numbers.

$$Deviation\ Ratio = \frac{\Delta f}{f_{max}} = \frac{5,000\ Hz}{3,000\ Hz} = \mathbf{1.666 \approx 1.67} \qquad (19.6)$$

Chapter 20

Bandwidth Calculations

We find it a bit unlikely that you will be spending a lot of time calculating the bandwidths of your various ham radio signals. We learn about the bandwidths of various modes very early in our ham careers and one of the first questions about any new mode that comes along is something like, "What kind of bandwidth are we looking at here?" The bandwidths of our various modes are – or at least should be – well known in the ham community.

This means our primary aim in this chapter is not just to teach the simple mathematics of calculating bandwidth (BW), but to deepen your understanding of the physics of modulation. Why is it that an SSB phone signal occupies about 3 kHz of bandwidth while an FM phone signal, carrying essentially the same information, is five times that width?

The calculations in this section have to do with the *necessary bandwidth* (BW_n) of various signals. Here's the official definition of that term from the NTIA; the US National Telecommunications and Information Administration.

> "Necessary Bandwidth: For a given class of emission, the width of the frequency band which is just sufficient to ensure the transmission of information at the rate and with the quality required under specified conditions."
>
> https://www.ntia.gov/sites/default/files/2023-11/j_2021_edition_rev_2023.pdf

"Necessary bandwidth" is a carefully chosen term. If we were to look at any given instant of most transmissions, the bandwidth occupied by the signal in that instant would be much less than the bandwidth necessary for the signal to work. The necessary bandwidth must be wide enough to accommodate the full range of frequencies that must be unoccupied by other signals for the mode to work properly. For instance, imagine a lower-sideband SSB RTTY signal which uses a 2,125 Hz tone to signify "mark" and a 2,295 Hz tone to signify "space." At the instant a mark is sent, we could argue that the bandwidth of the signal is effectively *zero*! After all, the carrier is suppressed as is the upper sideband, so the only thing being transmitted in that moment is an RF signal at 2,125 Hz below the carrier frequency.[1] However, the *necessary* bandwidth of an RTTY signal must take into account the 170 Hz spread between the tones, a bit extra for noise in the path, and yet another bit extra to account for the on/off switching of those tones. Once all that is accounted for, the typical bandwidth of an RTTY signal will be in the neighborhood of 250 Hz.

The fundamental law of modulation is that whenever we add information to a carrier we spread that carrier across the spectrum. It doesn't matter if that information is voice, the tones of an FT8 transmission, or even the dots and dashes of Morse Code. Put information on a carrier and you create bandwidth. Put on more information, create more bandwidth.

CW Bandwidth

If we're being perfectly correct in our language, transmitting Morse code by switching a carrier on and off is not "modulation" but "keying." For our purposes in this chapter, though, it is just another form of modulation.

Does CW occupy any bandwidth at all? Unmodulated carriers don't occupy any bandwidth (at least on paper). We're just switching an unmodulated carrier on and off, so how can there be bandwidth.

It is true that CW occupies *very little* bandwidth, but its bandwidth is not zero. According to that fundamental law of modulation, it can't be zero because we're putting information on the carrier. However, "it doesn't fit the equation" is not a very satisfying explanation.

The creation of bandwidth occurs in those moments of switching the carrier on and off. We might imagine that the carrier switches on and off

---

[1]This is a magic transmitter that completely suppresses the carrier and upper sideband.

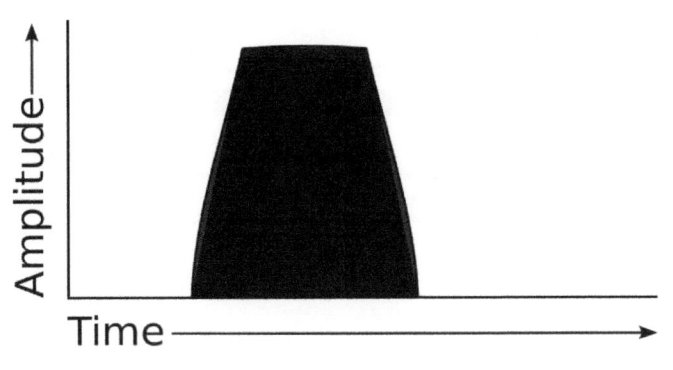

Figure 20.1: CW Waveform

instantaneously, but that's not only not the case, it cannot be the case; moving the carrier from 0%-on to 100%-on instantaneously would require an infinite amount of energy to be applied in zero time. Mathematicians would say we have a divide-by-zero problem. (Engineers would say we need a bigger power supply ...) Even if we could, somehow, set aside the laws of physics and create that "instant on" carrier, that leading edge of the waveform would be a square wave and square waves are loaded with harmonics. It would also be quite unpleasant for the listener's ears, since they would make very loud clicks with each character. The same would be true of an equally impossible "instant-off" trailing edge of the waveform. Those harmonics would occupy a *lot* of bandwidth.

The real waveform of a CW signal has a bit of slope to the leading and trailing edges, something along the lines of what you see in Figure 20.1. You will note that not even the top is a perfectly straight line in that (somewhat exaggerated) figure.

Each time that carrier switches on and each time it switches off, it mixes some frequency with the carrier and spreads it out to create bandwidth.

There's a handy rule of thumb that will allow you to almost instantly calculate the bandwidth of a CW signal based on the words-per-minute being sent. Just multiply the WPM by 4, and that's the approximate bandwidth. In the case of an operator sending 13 WPM, it works out like the formula below.

$$CW\ BW_n = WPM \times 4 = 13\,WPM \times 4 = \mathbf{52} \quad (20.1)$$

There are a few low-importance disclaimers that go with that formula,

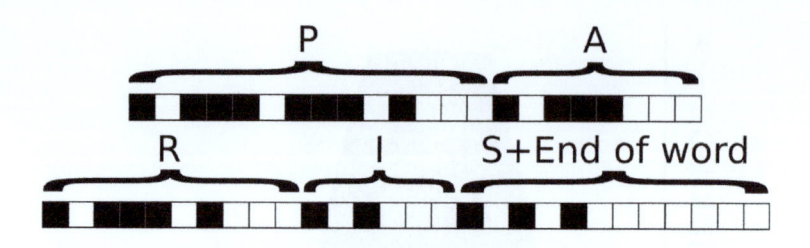

Figure 20.2: PARIS Consists of 50 Symbols

mostly having to do with the WPM figure. In English, "I" is a word. So is "floccinaucinihilipilification."[2] What are we counting as a word?

Generally, "words" in Morse code are five letters long. The "standard" word is PARIS. PARIS contains five letters that are transmitted using 50 symbols. A dot is a single symbol, as is the space between the dots and dashes of a single character. A dash is three dots long, as are the spaces between characters, so they count as three symbols. The space at the end of a word is seven symbols long. See Figure 20.2 for a graphic representation. This concept of symbols will become more important when we get to digital transmissions.

Amplitude Modulation Bandwidth

In the case of amplitude modulation the bandwidth is created by our old friend *heterodyning*; the mixing of two (or more) frequencies. When we modulate the carrier, we're mixing the carrier frequency with the modulating frequency. Once those signals are mixed we get the carrier frequency, the modulating frequency, a frequency equal to the carrier frequency plus the modulating frequency, and yet another equal to the carrier frequency minus the modulating frequency. We'd represent them mathematically as $f_{carrier}$, $f_{modulating}$, $f_{carrier} + f_{modulating}$, and $f_{carrier} - f_{modulating}$. That modulating frequency, $f_{modulating}$ is typically so far from the carrier frequency that we can safely ignore its presence.

Modern analog SSB transmitters use a balanced oscillator that does not output a carrier frequency. That leaves us with the upper and lower sidebands:

$$f_{carrier} + f_{modulating} = Upper\ Sideband$$

---

[2]The act of regarding something as unimportant or as worthless. "He regarded the word 'floccinaucinihilipilification' with great floccinaucinihilipilification."

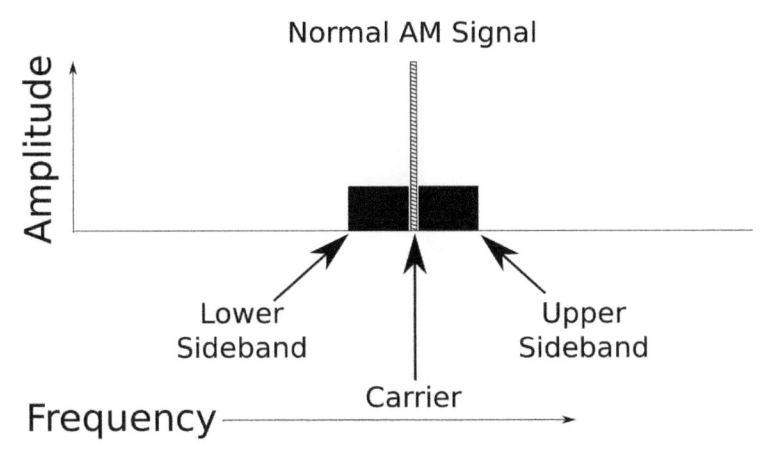

Figure 20.3: Amplitude Modulated Signal

$$f_{carrier} - f_{modulating} = Lower\ Sideband$$

A filter removes one of the sidebands and we are left with a single sideband signal.

You can see a visual representation of a "normal" amplitude modulated signal in Figure 20.3. You will note there is a considerable difference in amplitude between the carrier and the sidebands. Typically that difference is around 9 dB.

Figure 20.4 shows the same signal after the carrier and one of the sidebands is removed, leaving just the lower sideband. Now we can put all of the power of the transmitter into sending just that sideband. It's like we added 9 dB of gain to our antenna. Considering even many pricey HF tri-band antennas offer only around 7.5 dB of gain, that is a spectacular result.

If we were to feed an absolutely pure 1000 Hz test tone into an SSB transmitter set for upper sideband, it would produce a single signal 1000 Hz above the carrier frequency. On paper, at least, that signal has a zero bandwidth. That makes sense; it is carrying no information. Really, all we have done is moved the carrier 1000 Hz up in frequency. However, that signal had a beginning and will have an end. In those moments, the signal *will* occupy some bandwidth, just as CW occupies bandwidth.

Both Figures 20.3 and 20.4 show a signal that is being modulated by a microphone signal, or at least something similar. We'll assume it is a human voice. The full frequency range of a human speaking voice is from around 100 Hz to 8 kHz (most female voices start a little higher at

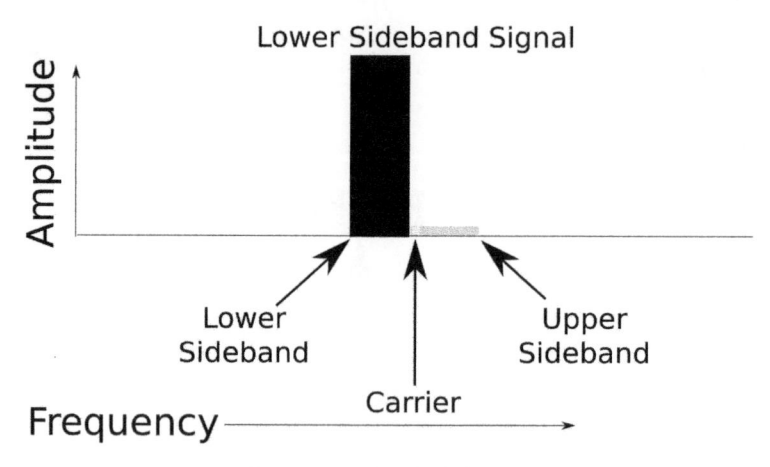

Figure 20.4: Single Sideband

150 to 250 Hz.) However, we don't need 8 kHz of bandwidth for a usable voice signal. Most of the intelligibility information of human speech is no higher than about 2,500 Hz, which fits neatly into the 3 kHz bandwidth of SSB.

If you listen to some speech through what's called a brickwall filter that eliminates almost everything above 2,500 Hz, it will sound very muffled and dull – almost like someone talking through a pillow. That's not how SSB sounds, though. How do we explain that? Bandwidth is not like a sack of potatoes. A five-pound sack of potatoes weighs five pounds; whatever the scale says, that's the answer, end of story. If you count every bit of signal that is created when you key the mic, you could make a case – a rather absurd case, but a case, nonetheless[3] – that all bandwidths are nearly infinite! Bandwidth is more like a five-pound sack of potatoes "plus some." How many is some? Chances are good a shopper's interpretation of "some" is different from the grocer's!

The communication authorities – that would be the grocer in this scenario – understand this, so they made a formal definition of bandwidth. The ITU defines bandwidth as the width of the frequency band where the mean power of the transmitted signal is 26 dB lower than the mean power within the band.[4] 26 dB is a lot, but it is not an infinite amount

---

[3]Your transmitter is generating harmonics and is also transmitting some level of broadband noise. Of course, at some point, all that disappears under atmospheric noise and the background radiation of the universe, but it is still present!

[4]Mean = average. Also note, there is a very small "conversion factor" to be applied to some emission types.

of dB, so some of that higher frequency information is getting through but it is at a low enough amplitude not to create meaningful interference with other signals.

AM Bandwidth Calculations

The calculation of bandwidth for any analog amplitude modulated signal is:

$$AM\ BW_n = K \times f_{max} \qquad (20.2)$$

BW_n is "necessary bandwidth" and f_{max} is the maximum frequency to be transmitted. K is a constant, reflecting the amount of distortion we're willing to tolerate and the quality of the signal path. For SSB phone signals 1.0 is a perfectly useful and standard value of K. Some SSB digital signals want a bit more quality; their K factor is typically 1.2. Double-sideband signals demand a K factor of 2 or better.

We can assign lower values of K if we are willing to tolerate lower quality and/or more errors. If we need more quality, that demands a higher K value. Standard definition fast-scan television typically uses a K factor of 4.2.

FM Bandwidth Calculations

We covered FM Deviation in Chapter 19. That modulation index tells us how far from the center frequency the signal will spread. That only accounts for one half of the total bandwidth, though.

In order to calculate FM bandwidth, we need to multiply the modulation index by 2. Mathematically, then, it is as you see in 20.3.

$$FM\ BW_n = 2 \times (\Delta f + f_{max}) \qquad (20.3)$$

In that formula, Δf is the carrier's greatest deviation, f_{max} is the highest frequency transmitted.

In terms of bandwidth efficiency, FM wins no prizes. Our "wideband" (US standard for hams) FM has a bandwidth of 13 to 16 kHz; roughly five times the bandwidth of an SSB signal.

Narrowband FM, with a Δf of just 2.5 kHz still occupies 11 to 12.5 kHz.

Perhaps we can take some small comfort from the commercial FM stations. With a Δf of 75 kHz they occupy about 200 kHz of bandwidth.

Digital Signal Bandwidths

Claude Shannon, Harry Nyquist, & Ralph Hartley

Claude Shannon, Harry Nyquist, and Ralph Hartley all worked at Bell Laboratories during the 1940's. Claude Shannon is often called "the father of information theory." Harry Nyquist worked in many areas and was said to have laid the groundwork for Shannon's later work in information theory. Ralph Hartley and Harry Nyquist were some twenty years older than Shannon, but there was a period when they all worked at the Labs at the same time. They are immortalized in the names of two of the fundamental equations of digital technology.

Each of them discovered remarkable things. (Among Ralph Hartley's inventions was the venerable Hartley Oscillator.) Harry Nyquist, in particular, seems to have been a catalyst for invention for both himself and those around him. At one point, Bell Labs wanted to know what qualities defined their most productive researchers in terms of patent production. After much research, it turned out that, "Workers with the most patents often shared lunch or breakfast with a Bell Labs electrical engineer named Harry Nyquist. It wasn't the case that Nyquist gave them specific ideas. Rather, as one scientist recalled, 'he drew people out, got them thinking.'"

It is impossible to transmit more information per second than twice the bandwidth of the signal. That is stated in the Shannon-Nyquist Theorem. In other words, if your signal is limited to 3 kHz of bandwidth, you cannot transmit more than 6,000 bits per second. Even that rate is conditional; the signal to noise level of the system (everything between the signal generator and the output of the receiver) must support that rate. In practice, the limit is always lower than $2 \times bandwidth$.

There is a formula to quantify the effects of noise in the system, known as the Shannon-Hartley Theorem.

$$C = BW\ log_2(S/N + 1)\ bits\ per\ second \qquad (20.4)$$

In that equation, C is capacity, BW is bandwidth, S is the average received signal power over the bandwidth measured in watts (or volts squared), and N is the average power of the noise and interference over the bandwidth, measured in watts (or volts squared). For this equation, we must use the linear power ratio of S/N, not logarithmic decibels.

The bottom line of that equation is something that probably counts as common sense for hams; if the noise is too loud, no information is transmitted. On the other hand, if you have 6dB (4×) signal-to-noise ratio, noise is no longer a problem because you have already hit the Nyquist Limit.

If you run the Shannon-Hartley equation with what we would consider normal signal-to-noise ratios, you'll see it predicting possible data rates far in excess of what's called the Nyquist Limit; $2 \times bandwidth$. No matter what the Shannon-Hartley equation says, the Nyquist Limit is still the real limit.[5]

Equation 20.6 shows how to calculate the base 2 logarithm of 1194.

$$log_2(1194) = \frac{log(1194)}{log(2)} = 10.2215 \qquad (20.6)$$

One interesting implication of the Shannon-Hartley theorem is that we can deal with higher noise levels by reducing the data rate of the transmission. This explains how modes like JT65 and FT8 can work so dependably in high noise situations. In fact, the theorem says we can still get data through even if the noise is louder than the signal! Let's imagine a horrible scenario where the signal-to-noise ratio is $\frac{1}{30}$. In other words, the noise is 30 times more powerful (14.7 dB) than the desired signal. We'll use a 3 kHz bandwidth.

$$C = B \times log_2(\frac{S}{N} + 1)\ bps = 3,000\ Hz \times log_2(\frac{1}{30} + 1) \approx 142\ bps \qquad (20.7)$$

According to Shannon-Hartley, even in a 1:30 signal-to-noise storm

---

[5]Should you feel the need, it is possible to calculate the log_y, the base y logarithm, of a number on your TI-30XS even though it only has a log_{10}. It does require an extra step. You divide the log_{10} of the number by $log_{10}(y)$.

$$log_y(x) = \frac{log_{10}(x)}{log_{10}(y)} \qquad (20.5)$$

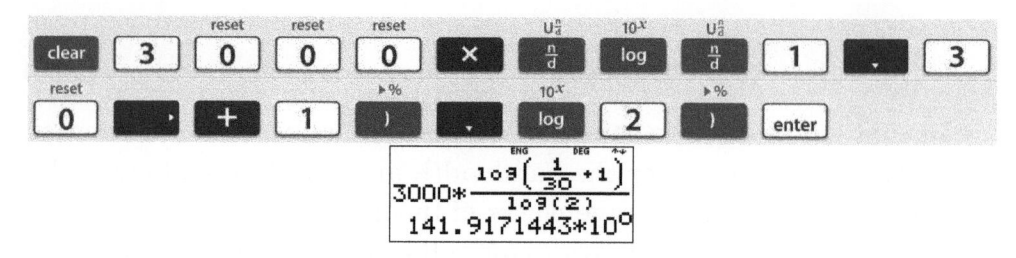

Figure 20.5: Keystrokes for Equation 20.7

(we should really call this a noise-to-signal ratio!) we can still get a maximum of 142 bits per second through the system. (That's excruciatingly slow, but it is more than zero.)

Calculating Digital Signal Bandwidths

HF digital modes work with some combination of audio tones. The differences among the frequencies of these tones is the *frequency shift* of the mode.

To calculate the bandwidth of data signals, we need to know the frequency shift, the baud rate, and that constant known as K. If you were sending RTTY with frequency shift keying using mark and space frequencies 170 Hz apart, the frequency shift would be 170 Hz.

The formula for calculating the bandwidth occupied by a signal with a given frequency shift and a given baud rate is:

$$bandwidth = (K \times shift) + B \qquad (20.8)$$

"B" is the baud rate of the transmission. Baud rate is the "symbols per second" rate.

Application

What is the necessary bandwidth (B_n) of an upper-sideband phone signal being modulated with frequencies up to 2,500 Hz?

The formula for the necessary bandwidth of any amplitude modulated signal is:

$$BW_n = K \times f_{max} \qquad (20.9)$$

For SSB phone signals we can use a K value of 1, so the necessary bandwidth of this signal is equal to the maximum frequency being transmitted: **2,500 Hz**.

What is the necessary bandwidth of a double-sideband phone signal being modulated with frequencies up to 2,500 Hz?

For a double-sideband signal we use the same formula as we used for single-sideband, but we use a K factor of 2.

$$BW_n = K \times f_{max} = 2 \times 2{,}500\, Hz = \mathbf{5{,}000\, Hz} \tag{20.10}$$

What is the necessary bandwidth of a frequency modulated signal with a frequency deviation of 15 kHz and a maximum transmitted frequency of 6 kHz?

For FM, $B_n = 2 \times (\Delta f + f_{max})$, so:

$$BW_n = 2 \times (\Delta f + f_{max}) = 2 \times (15\,kHz + 6\,kHz) = 2 \times 21\,kHz = \mathbf{42\,kHz} \tag{20.11}$$

What is the necessary bandwidth of a 170-hertz shift, 300-baud ASCII transmission?

We're given the frequency shift, the baud rate, and we know that K is 1.2, since this isn't CW and it isn't AM — neither of those shift frequency. It doesn't matter a bit that it is an ASCII transmission. We have everything we need to calculate this bandwidth.

$$BW_n = (K \times shift) + baud = (1.2 \times 170) + 300 = 504\,Hz \approx \mathbf{0.5\,kHz} \tag{20.12}$$

It's always valuable to be able to evaluate whether a possible answer to a question makes sense. So let's look at this problem a little differently — and you can apply this line of thinking to similar problems as well.

The basics of Information Theory tell us that bandwidth must be *at least* equal to the baud rate of information being transmitted. So if we're transmitting 300 baud, we need a bare minimum of 300 Hz of bandwidth. That's one of those ideal numbers, so we'll never really fit 300 baud into 300 Hz of bandwidth, but we might get in the ballpark, at least.

Look at the possible answers to this question. 0.1 kHz *can't* be right. There's just not enough room! It's like trying to pack twenty bologna sandwiches into little Jimmy's lunch sack. 0.3 kHz isn't possible either.

1.0 kHz seems a little excessive. That leaves **0.5 kHz** as the answer that seems about right.

What is the necessary bandwidth of a 4,800-Hz frequency shift, 9,600-baud ASCII FM transmission?

$$BW_n = (K \times shift) + baud = (1.2 \times 4,800) + 9,600 = \mathbf{15.36\,kHz} \qquad (20.13)$$

According to the Shannon-Hartley Theorem, what is the capacity (C), in bits, of a radio link with a bandwidth (B) of 2,600 Hz and a signal to noise ratio of $2\,\mu V : 1\,\mu V$?

If this was test, this would definitely be an extra credit question. Here's how it works out:

$$C = B \times log_2(\frac{S}{N} + 1) = 2,600\,Hz \times log_2(\frac{2}{1} + 1) \approx 4121\,bps \qquad (20.14)$$

According to the Nyquist Limit, is this bit rate possible?

Chapter 21

ERP, EIRP, Link Budget & Link Margin

The various agencies who regulate amateur radio around the world specify different ways of measuring the power we are allowed to transmit.

The simplest is PEP, Peak Envelope Power. If you wanted to calculate it you'd multiply your transmitter's output voltage by the output amperage but we rather doubt anyone ever does that. Your PEP is whatever the wattmeter on your radio or the separate wattmeter you have attached to the radio says the output is. If you have a 100-watt radio and the power is cranked up all the way, your PEP is 100 watts.

ERP and EIRP both take into account any losses or gains in the transmission system. ERP, Effective Radiated Power, is the strength of the electromagnetic field coming off your antenna in the forward direction compared to the strength of the field coming from a half-wave dipole fed by the same transmitter and feed line. EIRP is similar, but compares the output to an isotropic radiator. That's the imaginary "perfect" antenna with zero loss and zero gain.

Think of it this way. Let's say you have an antenna with 10 dBd of gain in the forward direction. That's 10 dB of gain compared to a "d"; a dipole. For the moment, let's also say that there are no other gains or losses in the system. You feed that antenna with 100 watts. Since 10 dBd is equal to a 10 times gain in radiated power, your ERP in this imaginary system would be 1,000 watts.

It's easy enough to calculate the overall gain or loss of the transmission system. You just add up all the gains and subtract all the losses – or add the losses as negative numbers.

$$dB_{Total\ Gain\ or\ Loss} = dB_{Gain_1} + dB_{Gain_2} + dB_{Gain_3} \ldots dB_{Gain_n} \qquad (21.1)$$

Enter all losses as negative numbers, using the [(-)] key. That will give us the total dB of gain or loss. Then we need to convert that to a power factor, another way to say "gain", because eventually we'll need to multiply the transmitter output by that power factor.

$$Power\ Factor = 10^{\frac{dB_{Total\ Gain}}{10}} \qquad (21.2)$$

Raise 10 to the power of the dB of gain divided by ten to get the power factor. Then we multiply the transmitter power by the power factor to get the Effective Radiated Power.

If you have a net loss instead of a gain, you'll enter a negative exponent in the power factor calculation. For instance, if you had a 3 dB loss, you'd enter $10^{\frac{-3}{10}}$ or $10^{0.3}$. Once you have the power factor, the ERP calculation is simply the transmitter's output power multipled by the power factor.

$$ERP = Transmitter\ Output\ Power \times Power\ Factor \qquad (21.3)$$

Application

What is the effective radiated power (ERP) (relative to a dipole) of a repeater station with 150 watts transmitter power output, 2 dB feed line loss, 2.2 dB duplexer loss, and 7 dBd antenna gain?

We'll show you how to calculate this precisely, but we can "eyeball estimate" our way to a fairly accurate answer. Here's the estimation method. First, we need to know the overall gain (or loss) of this system. That's as simple as adding up all the gains and losses.

$$dB_{Total\ Gain\ or\ Loss} = -2\,dB + -2.2\,dB + 7\,dBd = 2.8\,dB\ (of\ gain) \qquad (21.4)$$

We know 3 dB equals a doubling of power. 2.8 is pretty close to 3, so the answer is going to be just a little less than 150 watts times 2; 300 watts. Now, when we do our calculation, we'll have a good sense of whether our answer is correct or not.

Here's the real calculation. First we calculate the multiplier or "power factor" for 2.8 dB. To do that, it's 10 raised to the power of the dB divided by 10. $10^{\frac{2.8}{10}}$. That's 1.905×; the ERP will be 1.905 times the transmitter output power. Then we multiply 150 watts by 1.905.

$$Power\ Factor = 10^{\frac{2.8\,dB}{10}} = 1.905\times \qquad (21.5)$$

$$ERP = Power\ Factor \times Power_{Out} = 1.905 \times 150\,W \approx \mathbf{286\,W} \qquad (21.6)$$

Whoa, but wait! That question asked about power "relative to a dipole." Do we have to do something about that?

Nope. Notice the gain of the antenna is given in **dBd**. The little "d" on the end of dBd tells us that's already "dB of gain relative to a dipole" so it's all handled in the setup of the question.

In practice, the gain of many antennas is advertised as "dBi." That's gain relative to an isotropic antenna. If you need your ERP relative to a dipole, you'll subtract 2.15 dB from the dBi figure to convert it to dBd. In most countries, power on the 60-meter band is limited to 100 watts **dBd**, so be sure you're using the right gain figure.

What is the effective isotropic radiated power (EIRP) of a station with 200 watts transmitter power output, 2 dB feed line loss, 2.8 dB duplexer loss, 1.2 dB circulator loss, and 4.85 dBd antenna gain?

This question asks about *EIRP*, not *ERP*. ERP is *effective radiated power*, EIRP is *effective isotropic radiated power*. EIRP[1] is "radiated power compared to an isotropic antenna" while ERP is "radiated power compared to a dipole."

Because we have been given the gain in dBd and we want to know EIRP, we need to first convert dBd to dBi by adding 2.15 dB to the dBd figure for a dBi figure of 7 dBi.

$$dB_{Total} = (-2\,dB) + (-2.8\,dB) + (-1.2\,dB) + 7\,dB = 1\,dBi\ Gain \qquad (21.7)$$

$$Power\ Factor = 10^{\frac{1\,dB}{10}} = 1.258\times \qquad (21.8)$$

$$EIRP = Power\ Factor \times Power_{Out} = 1.258 \times 200\,W \approx \mathbf{252\,W} \qquad (21.9)$$

Link Margin & Link Budget

In life outside the ham exam room, you might never calculate a link budget and/or link margin, but the concepts behind them are fundamental to all radio communications. They're especially important to understand when you are trying to figure out why some communications link is not working.

---

[1]EIRP is the measurement used in US regulations for the 630-meter and 2200-meter bands.

Link margin and link budget are concepts that apply to signal paths. We all understand the idea that we can't hear a signal that's too weak to hear. Link margin and link budget quantify that idea by evaluating each point of possible gain or loss along that signal's path to our ears.

Link budget is "how much of the signal from transmitting system A actually arrived at receiving system B?" Here's the formula. Calculating this is just a longer version of what you did to calculate ERP and EIRP.

$$Link\ Budget = P_{RX} = P_{TX} + G_{TX} - L_{TX} - L_{FS} - L_M + G_{RX} - L_{RX} \quad (21.10)$$

P_{RX} = Received power; the amount of power received at the input to the receiver.

P_{TX} = Transmitter output power.

G_{TX} = Transmitter antenna gain.

L_{TX} = Transmitter losses. (Feed line, connectors, SWR)

L_{FS} = Path loss or "Free space loss." The effect of distance and anything else in the signal path on the signal.

L_M = Miscellaneous loses. (Polarization mismatch, etc.)

G_{RX} = Receiving antenna gain.

L_{RX} = Receiver losses. (Feed line, connectors, etc.)

All values in dB except the transmitter output power which is expressed in dBm; *decibels compared to 1 millivolt.*

Link margin is **the difference between received power level and minimum required signal level at the input to the receiver.** Another way to put it is that it is the difference between the minimum required signal level at the input to the receiver and the link budget. For instance, our IC-7300 radio in SSB mode needs to see 0.16 μV (microvolts) of signal – 0.16 μV of link budget – in order to hear a station. If what's arriving at the receiver input is less than 0.16 μV, we have a negative link margin and awkward silence.

Application

What is the link margin in a system with a transmit power level of 10 W (+40 dBm), a system antenna gain of 10 dBi, a cable loss of 3 dB, a path loss of 136 dB, a receiver minimum discernable signal of -103 dBm, and a required signal-to-noise ratio of 6 dB?

To get the link margin we first need to know the link budget of this system.

$$Link\ Budget = P_{RX} = +40\,dBm + 10\,dBi - 3\,dB - 136\,dB = -89\,dBm \quad (21.11)$$

This tells us the signal is arriving at the receiver at -89 dBm; 89 dB below one millivolt. All we need to know now is what the receiver wants to see. We're told it has a minimum discernible signal of -103 dBm. Then they throw in that it has a required signal-to-noise ratio of 6 dB. ("Discernible" does not equal "readable.") What do we do with the 6 dB? That figure tells us we need to *add* 6 dB to the minimum discernible signal figure of -103 dB.

$$Signal\ Strength = -103\,dBm + 6\,dB = -97\,dBm \quad (21.12)$$

We need at least -97 dBm to make this receiver work. We have -89 dBm, so:

$$Link\ Margin = -89\,dBm - (-97\,dBm) = \mathbf{+8\,dBm} \quad (21.13)$$

We have a link margin of **+8 dB**.

We have never seen a power meter or radio that gives a direct readout in dBm (power relative to a milliwatt.) How do we convert output power to dB for these calculations?

By now you can probably guess that logarithms are involved. Here's the formula.

$$Power(dBm) = 10 \times log_{10}(1000 \times P) \quad (21.14)$$

Power in *dBm* is 10 times the logarithm of 1000 times the output power in watts.

What power level does a receiver minimum discernible signal of -100 dBm represent?

This is a conversion from dBm to watts; the reverse of what the formula above accomplishes. Here's the formula:

$$P_w = \frac{10^{\frac{dBm}{10}}}{1000} \quad (21.15)$$

When we plug -100 dBM into that equation, we get what you see in Equation 21.16.

$$P_w = \frac{10^{\frac{-100 dBm}{10}}}{1000} = 100 \times 10^{-15}\,W = .1 \times 10^{-12}\,W = 0.1\,pW \quad (21.16)$$

In strict engineering notation, the correct answer would be 100×10^{-15}; 100 *femtowatts*. "Femtowatts" is not a value that is commonly used in amateur radio, so we have converted to picowatts which is at least a tiny bit more common in our vocabularies. In strict scientific notation, the answer is 1×10^{-13}; there's no metric prefix for 10^{-13}. That's why there's no way to make the TI-30XS have the answer come out to 0.1×10^{-12} so we did the final conversion to **0.1 picowatts** (pW) manually.

The "m" in dBm stands for milliwatts, so we're comparing some received signal strength to the strength we'd get if the signal's power at the receiver was 1 milliwatt.

The question says the signal is -100 dBm; 100 dB lower than that 1 milliwatt signal. We need to know what power factor -100 dB represents. To do that, we use this formula;

$$Power\ factor = 10^{\frac{dB}{10}} = 10^{\frac{-100\,dB}{10}} = 100 \times 10^{-12} = 0.1\,pW \qquad (21.17)$$

Chapter 22

Transmission Lines

We can create a transmission line from two lengths of wire with some sort of insulator separating the two lengths. There are various variations on that theme, known as window line and ladder line. They're very low loss transmission lines compared to coaxial cable. They're even a very good impedance match to certain types of antenna, eliminating the need for any sort of impedance matching device between the line and the antenna.

There are problems with those lines, though. First, they radiate RF. That RF can interfere with equipment, and can couple to nearby objects altering the impedance of the transmission line. Those lines are also rather intolerant of being bent or twisted – more changes in the impedance.

Coaxial cable keeps the RF (mostly) inside the cable. That is one of the main reasons to use coaxial cable.

Figure 22.1: Window Line

Coax does not act quite like the wiring in your house. In your house wiring, electrons are shuttling back and forth in time with the frequency of the household current. In coaxial cable, the electrons are pushed together then pulled apart as the wave passes. This means the coax isn't carrying an electrical current in the way we usually think of that; it is propagating an electromagnetic wave.

If you have ever had occasion to attach a connector to a coaxial cable, you know that they do not appear to be complicated devices. There's a center conductor of some sort, a layer of insulating (*dielectric*) material, one or two layers of shielding and some sort of insulating jacket around the various parts. Compare a coaxial cable to a clock or a steam engine and the coaxial cable seems almost childishly simple. Yet it is quite possible for an electrical engineer to spend a substantial part of a career specializing in the study of transmission lines in general which almost always means coaxial cable in particular.

Coaxial cable is a bit of an oddity among inventions in that it was manufactured in great quantity before it was, technically speaking, invented! The first transatlantic telegraph cable was opened in 1858. Oliver Heaviside would not describe the operation of coaxial cable until 1880, the year he filed a patent for its invention.

Figure 22.2: Cable Manufacturing *circa* 1858

In part because the 1858 understanding of coaxial cable characteristics was more than a little lacking, in part because the cable's manufacture and installation were rushed, and in large part due to some "operator error," the first transatlantic cable was a short-lived mess. The cable was severely under-engineered. That was in large part due to the efforts of one Dr. Edward Whitehouse the Atlantic Telegraph Company's "chief electrician." Dr. Whitehouse's doctorate was in medicine; he was a surgeon with a hobby of dabbling in electrical experiments. He did, however, speak the words best understood by the head of the cable company; "I can save you a fortune!" His "brilliant" plan was to make the cable thinner.

The cable did manage to transmit one message, from Queen Victoria to President James Buchanan: "Directors of Atlantic Telegraph Company, Great Britain, to Directors in America: Europe and America are united by telegraph. Glory to God in the highest; on earth peace, good will towards men." Thanks to the cable's woefully poor design it had a miniscule *velocity factor*. In other words, it slowed the signals passing through it, and not just a little bit. That message is about 40 "Morse code words" long. It would take a competent telegraph operator a minute or less to transmit. The message took 16 *minutes* to receive.[1] The dots and dashes had been smeared out over time by the velocity factor of the cable. The return message from the US side took even longer. The more letters they attempted to send, the slower the cable became. This additional slowing seems to have resulted from the mis-designed cable being a colossal capacitor, refusing to accept more charge.

---

[1] It appears it actually took *days* to receive from the time it was sent, since it went through a relay station in Newfoundland which experienced its own set of problems.

For a company whose whole business plan was to charge for sending messages by the word, this was a disaster. Whitehouse sprang into action and promptly did the worst possible thing! He thought he could overcome the *velocity factor*, known in those days as *retardation*, by increasing the voltage applied to the cable. Apparently this was over the heated objections of some more knowledgeable members of the team, especially William Thomson whom you might know by his later title, Lord Kelvin. Poorly designed from the outset, mishandled at every step from its hasty manufacture to its transport to its placement under the ocean, the cable – only $\frac{5}{8}$ of an inch in diameter – quickly shorted out under the strain of up to 2,000 volts and that was the end of that venture.

Happily, we have learned volumes about how coaxial cable functions since that doomed attempt, and today the globe is spanned by millions of miles of coaxial cable.

The "volumes" we have learned consist mostly of mathematics. We won't delve into the many formulas that govern coaxial cable construction. We doubt you will be constructing your own coaxial cable, after all, since any type you might want is readily available. We will look at a couple of the foundational formulas, not for calculation purposes, but, we hope, to illuminate some aspects of these ubiquitous transmission lines.

There are practical, mathematically based transmission line problems we occasionally need to solve, but for most of those there are Smith Charts. Smith Charts are a tool you will meet in Chapter 23.

Anatomy & Physiology of a Coaxial Cable

Most hams are familiar with coaxial cable. Figure 22.3 offers a quick look at what is inside that expensive "signal hose."

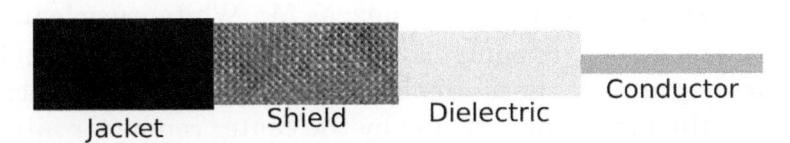

Figure 22.3: Internal Structure of Coaxial Cable

These days the jacket is typically some sort of plastic. In that first transatlantic cable (and for many years thereafter) the insulating jacket and dielectric were gutta percha; the boiled sap of the gutta tree.[2] The shield in flexible coaxial cable is braided copper wire. As a coaxial

---

[2]Gutta percha is not quite 100% obsolete. If you've ever enjoyed (?) a root canal treatment, the root of that tooth is probably packed with gutta percha.

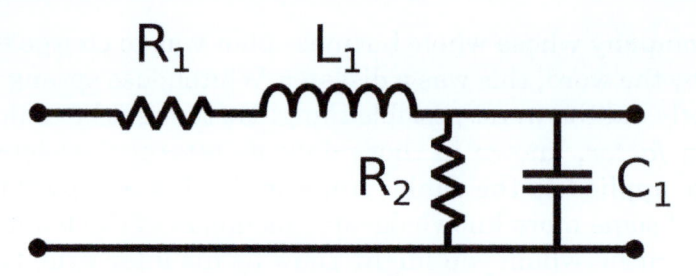

Figure 22.4: Equivalent Circuit of Coaxial Cable

cable bends, the braid inevitably develops gaps, so it is often augmented by either another layer of braid or by a layer of foil wrapped under the braided shield. The composition of the dielectric is critical to the performance of the cable. Dielectric materials can be solid polyethylene plastic, foamed polyethylene, *FEP* (Fluorinated Ethylene Propylene) which is a type of Teflon, *FFEP* (Foamed Fluorinated Ethylene Propylene), or air. Each has a different *dielectric constant* which is the main factor in the velocity factor of the cable as well as most of the other performance parameters.

The conductor can be solid or stranded and can be copper or copper-plated steel. If the cable uses a steel core, it cannot be repeatedly flexed without risking a failure due to metal fatigue. [3]

Beyond the cable's electrical characteristics, there are many other factors to consider when selecting coaxial cable. They include whether the cable will be installed indoors or outdoors, whether it will above-ground or below, how much it must be bent and whether it will be bent once or will flex. If installed indoors, will it be in a *plenum space*, like the space above a suspended ceiling? *Rated for Plenum Use* is yet another special rating for a cable. Finally, maximum power handling capacity may be important in some applications, as Mr. Whitehouse learned.

Figure 22.4 shows an equivalent circuit of a coaxial cable. In that schematic diagram, R_1 is the resistance of the center conductor and shield, L_1 is the inductance created by the center conductor interacting with the shield, C_1 is the capacitance created by the proximity of the two conductors, and R_2 is the resistance of the dielectric material – typically very high, but not infinite. Put another way, there is some amount of conductance between the center conductor and the shield. [4]

---

[3] At radio frequencies, skin effect will cause most of the signal to be carried in the copper plating, so the additional resistance of the steel has a minimal effect on performance at higher frequencies. Steel core cable may not be optimum for lower HF frequencies.

[4] Note: In formulas, the value we have shown as R_2 is typically shown as G; the siemens of conductance between the shield and center conductor.

Clearly, this is a resonant circuit, and just as clearly, the properties of this circuit are determined by the L, C, and R in the circuit. This resonant character is why we refer to coaxial cable as *transmission line* rather than just a pair of conductors.

Think back to those phasor diagrams, those impedance triangles. If the L and C are equal, then the "triangle" becomes a straight line. Essentially, that's what happens in an ideal coaxial cable. The reactances cancel each other and we're left with 50 Ω of resistance.

Why doesn't the resistance go up with length? Remember, there's some conductance between the center conductor and the shield. As the schematic shows, that acts like a parallel resistor. In this case, we'd call it a *shunt* resistor, but for calculation purposes it acts like a parallel resistance.

As we add more length the cable, it's like we're stringing more resistors in series, but we're also connecting more resistors in parallel, so the resistance value remains the same.

That's not to say there are no losses in a coaxial cable. Losses increase with length *and* with frequency. There are three types of loss in a coaxial cable.

1. Resistive loss due to the inherent resistance of the conductors. Resistive loss increases with frequency as the signal crawls onto the outer layers of the conductor, reducing the conductance available.

2. Dielectric loss. This is signal energy lost to heat as the charge passes through the dielectric material. This type of loss also increases with frequency. (In the schematic, this is represented by the shunt resistor, R_2.)

3. Radiation loss. No shield is perfect, so some of the signal energy leaks out into space, never to return. This is usually a very minor part of the overall loss.

All the values change with frequency but, ideally, remain in harmony throughout the cable's useful range. This is how a five-foot length of RG8X can have the same impedance as a fifty-foot length.

Mathematically, the simple equation for a coaxial cable's impedance is:

$$Z = \sqrt{\frac{L}{C}} \qquad (22.1)$$

That equation assumes you somehow know the L and C values to plug into the equation so it isn't very useful but it does show the components of the cable's impedance, and resistance is not part of it.

A more complex equation allows us to use the key measurements of a transmission line to determine the impedance. For twin-conductor transmission lines such as ladder line, the equation is:

$$Z = \frac{276}{\sqrt{k}} log_{10} \frac{d}{r} \qquad (22.2)$$

In that equation, Z is the characteristic impedance of the line, 276 is a constant, k ia the *relative permittivity of insulation between the conductors*, d is the distance between the conductors, and r is the radius of the conductors.

276? Where'd they get 276? 276 is a value that helps make the conversion from one of Heaviside's Telegrapher's Equations into something us mere mortals can use.

For coax, the equation is slightly different, as you can see in Equation 22.3.

$$Z = \frac{138}{\sqrt{k}} log_{10}(\frac{d_1}{d_2}) \qquad (22.3)$$

In that equation, Z is still the characteristic impedance of the coaxial cable, 138 is the constant, k is the relative permittivity of the dielectric material, d_1 is the inside diameter of the outer conductor (the shield) and d_2 is the outside diameter of the center conductor.

The units used to measure the diameters are not important, since we're only concerned with the ratio of the two measurements.

The relative permittivity value is something you would look up. A vacuum has a relative permittivity of 1. Air as a relative permittivity so close to 1 it really makes no difference unless you're working to tolerances of 1/10,000th.

The various Teflons all have relative permittivities very close to 2, so there is not a huge range of values that are likely for k. After all, the square root of 1 is 1 and the square root of 2 is 1.414, so that first fraction in the equation works out to something between 97.58 ($\frac{138}{1.414}$) and 138 ($\frac{138}{1}$).

Because a piece of coax is a resonant circuit, we can use it as an impedance matching device or even as a filter, and an almost zero-loss one at that. The catch is that while coaxial cable itself has enormous bandwidth, the Q of a coax impedance match or filter is sky high so they are more or less single-frequency devices. They're handy in a repeater or auxiliary station installation, useless in a typical station that needs to operate on multiple frequencies.

Applications

Transmission Lines for Impedance Matching

Among the many tools we have for impedance matching is something as simple as a carefully cut length of coaxial cable.

Which of these transmission line impedances would be suitable for constructing a quarter-wave Q-section for matching a 100-ohm feed point impedance to a 50-ohm transmission line?

A *Q section* is an impedance matching device constructed from nothing more than a length of coaxial cable.

There are two ways to approach this question. One is, memorize "75 ohms is halfway between 50 and 100 ohms."

The other way is with the actual formula for an impedance matching transformer, which is what that 1/4 wavelength piece of 75-ohm coax will become. Specifically, it becomes what is known as a synchronous transformer.

The formula is related to the transformer formula you saw in Chapter 17. This time, though, we're not concerned with the "turns" in a transformer's primary and secondary inductors; we just want to know what impedance to stick between our mismatched impedances.

$$Z_{Transformer} = \sqrt{Z_1 \times Z_2} \tag{22.4}$$

We just multiply the two mismatched impedances together, take the square root of the product, and that's the impedance we want for the matching transformer.

$$Z = \sqrt{Z_1 \times Z_2} = \sqrt{100\,\Omega \times 50\,\Omega} = \sqrt{5000\,\Omega} = 70.71\,\Omega \approx 75\,\Omega \tag{22.5}$$

You can see we came up with 70.71 ohms, but the ham radio store was fresh out of 70.71 ohm coax, so 75 ohms is, as we say, close enough for radio!

All this raises a more basic question: Why does a coaxial cable have an impedance at all and what makes one cable a 50 Ω cable and another a 75 Ω cable?

Physical size makes a huge difference in the values of just about any electronic component. Resistors, capacitors, inductors, and antennas all change value as they change size. That's not true of the impedance of a coaxial cable. A 5-foot length of RG-8X and a 100-foot length of RG-8X have (almost) the same impedance.

Here's an equivalent circuit schematic for a piece of coaxial cable.

Velocity Factor

Radio waves don't *always* travel at the speed of light. They only really do that in a vacuum, out in space, but for most purposes, "speed of light" is close enough. Not when it comes to transmission lines, though, when physical dimensions relative to wavelength get very important.

The velocity factor of a coaxial cable is a function of the inductance and capacitance of the cable. In the most general case, it is a function of the *relative permeability* and the *relative permittivity* of the cable. Permeability has to do with inductance while permittivity has to do with capacitance.[5]

Equation 22.6 shows the formal general formula for velocity factor.

$$VF = \frac{1}{\sqrt{\mu_r \epsilon_r}} \qquad (22.6)$$

μ_r is the relative permeability, ϵ_r is the relative permittivity. When we get to Maxwell's Equations in the last chapter you will see that that VF formula is almost identical to a formula for the speed of light.

Permeability and permittivity are not values we use a lot in amateur radio. Equation 22.7 is a formula that is more practical for us. Even though it is for an ideal, lossless coaxial cable, it clearly reveals the factors at work.

$$VF = \frac{1}{c_0 \sqrt{L'C'}} \qquad (22.7)$$

---

[5]Permeability is a measure of how easily magnetic lines of force can pass through a material. Permittivity is a measure of a material's ability to store energy in an electric field.

In the equation c_0 is the speed of light in a vacuum, and L' is the *distributed inductance* and C' is the capacitance of the cable.

Distributed is a term you will come across often in transmission line studies. It simply means "the total of all the (capacitances or inductances) in the line." A related term is *per-unit-length*; that's referring to the capacitance or inductance at a single spot in the line. In the case of this formula, we're looking for the per-unit-length value of capacitance.

As you can see, R is rather notably absent from that equation. Had poor old Whitehouse known that, he would have known that applying 2,000 volts to that poor transatlantic cable would accomplish nothing except a brief warming of a few molecules of water at the bottom of the Atlantic.

In a typical ham radio coaxial cable with a solid polyethylene dielectric – that white core inside, say, RG-58 – the wave slows down to only 66% of the speed of light. Another way to say that is that RG-58 has a *velocity factor* of 0.66.

If you are studying for a ham exam, there's a chance you will be asked to calculate something like the physical length of a piece of, say, polyethylene dielectric coax with an electrical length of 1/2 wavelength. You need to know the velocity factor for that type of coax then apply it to a wavelength calculation, just like the calculation you'd do to figure out the length of a 1/2 wavelength antenna. The velocity factor is the ratio of the actual speed of propagation through the cable to the speed of light in a vacuum.

Let's imagine that for some reason you need to cut a piece of coax that is $\frac{1}{4}$ wavelength long at 14.075 MHz. The coax you have has a polyethylene dielectric, so it has a velocity factor of 0.66.

First, let's determine the wavelength (λ) of a 14.075 MHz signal in meters.

$$\lambda_m = \frac{300}{f_{MHz}} = \frac{300}{14.075 MHz} = 21.31\,m \qquad (22.8)$$

To get the length of $\frac{1}{4}$ wavelength we'll divide 21.31 meters by 4.

$$\frac{1}{4}\lambda = \frac{\lambda}{4} = \frac{21.31}{4} = 5.33\,m \qquad (22.9)$$

If we cut that coax to a length of 5.33 meters it will be too long. We still need to apply that velocity factor of 0.66.

$$Length = Physical\,Length \times Velocity\,Factor = 5.33\,m \times 0.66 = 3.5\,m \quad (22.10)$$

We'll want to cut that coax about 3.5 meters long. In practice, of course, we'll cut it just a bit long then trim to adjust.

Chapter 23

The Smith Chart

When we start calculating things like, "If the impedance of my antenna is 50 ohms at frequency X, what is the impedance at frequency Y," and "what impedance does a ¼ wavelength long piece of transmission line present if the far end is shorted?" or even, "What components should I use to build a matching network?", we quickly find the maths get, to use a very technical maths term, "hairy." That's one reason we have Smith Charts.

While this book has been, until now, about how to calculate various values, this chapter is about a tool for avoiding a lot of calculation.

Smith Charts are a way to use pencil and paper to work out what would otherwise be quite complex problems. Today, you're unlikely to purchase a pad of Smith Charts, sharpen up your pencil, and get right to work. Aside from long-suffering Electrical Engineering students, there's just no need for that. We have electronic calculators and computers to handle that sort of thing, and handle it faster with more accuracy. However, the Smith Chart is still a very good way of graphically representing complex impedances. You'll find a simplified Smith Chart incorporated in the display screens of vector network analyzers. Even the very popular nanoVNA, which often sells for under $50, offers a Smith Chart representation of the network it is measuring. With all that in mind, we will focus on what the chart represents rather than all the techniques of using it. Should you wish to know those techniques, there are several texts available, including Phillip Smith's book, *Electronic Applications of the Smith Chart: In Waveguide, Circuit, and Component Analysis (Electromagnetic Waves)* (264 pages!) (https://amzn.to/47OWWyh) which presently sells for about $200. (Less expensive alternatives including various ebooks and videos on YouTube are readily available.)

Phillip Smith was a researcher at Bell Labs, where he went to work in 1928. He sounds like quite an interesting guy. When he was going through Electrical Engineering school he commuted in a reconstructed Model T, then later on his Harley-Davidson motorcycle. He was a ham radio operator – call sign 1ANB, and a private pilot. He did some of the early development work on directional antenna arrays for commercial AM radio stations. He had a special interest in transmission lines, and matching them to antennas. This was quite the tedious problem in those days, involving numerous physically challenging measurements. He just knew there had to be a better way.

The Smith Chart was the better way. By 1936, or thereabouts, he had developed the beginnings of what we now know as the Smith Chart as an easy way to see what would be the results of vast volumes of calculations on one chart.

Around 1940 came the real breakthrough for the chart. Smith tossed his rectangular coordinate graph paper out the window, then sent his polar coordinate graph paper out right behind it.[1] Instead he plotted all those values on graph paper with axes that were a series of circles offset from each other – and that let him represent resistance, capacitive reactance, inductive reactance, impedance, SWR, and their interrelationships all on one chart. Solving complex transmission line problems went from hours of slide rule calculations – no computers in those days! – to tracing a few lines on a chart, and maybe using a compass to draw a circle.

Smith went on to contribute to the development of radar, and you can thank him for high-power coaxial lines and the adjustable stub tuner, among other things.

Okay, here comes the only scary part of the chapter – your first look at an actual Smith Chart. Don't panic – remember, it makes things *easier*!

The chart used by most network vector analyzers is a vastly simplified version of the illustration in Figure 23.1 – but we want you to see what the real graph paper looks like. At the size we can reproduce, the numbers on this one probably won't be legible – don't worry about that. For now, the important thing to learn is the parts of the chart and what the chart represents. If you would like a full-size chart you can download a vector graphics file of one at

https://upload.wikimedia.org/wikipedia/commons/2/2a/Smith_chart.svg

If you print that file, it looks best on oversize paper such as 11x14 or A3.

Assuming you didn't just run out of the room screaming, let's take

---

[1]The Smith Chart is, in fact, a highly modified polar coordinate chart but is barely recognizable as such.

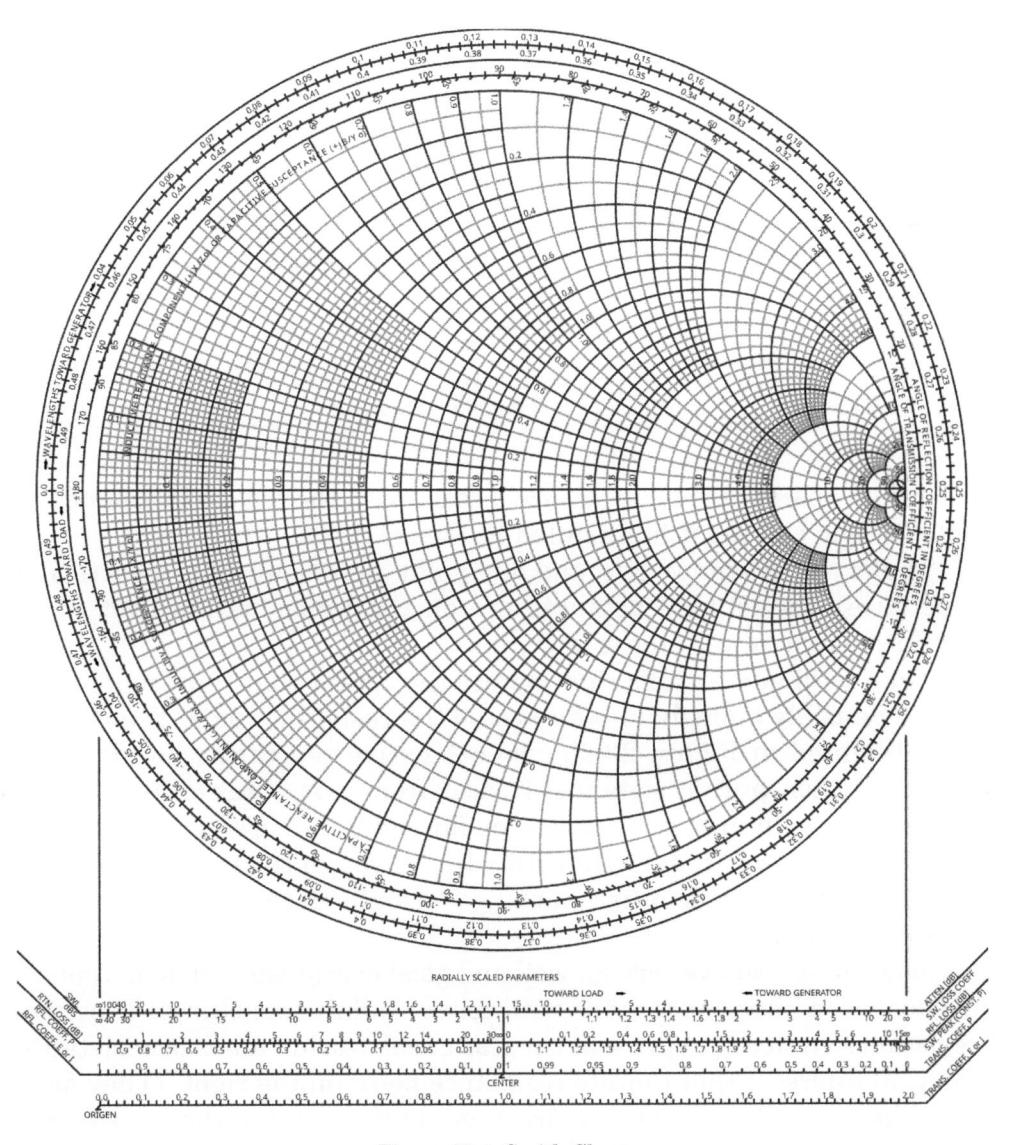

Figure 23.1: Smith Chart

(a) Reactance Axis　　　　　　　　　　(b) Resistance Axis

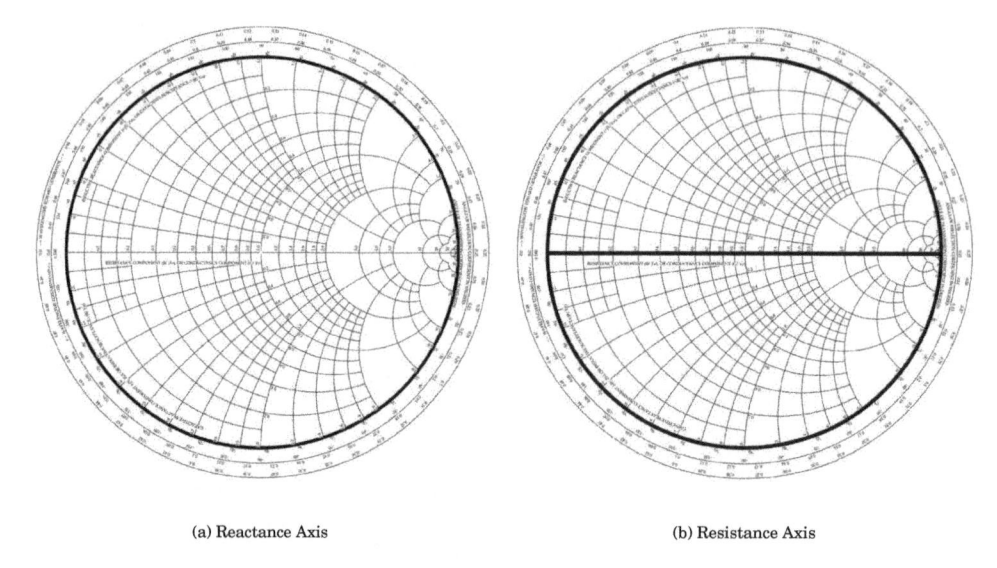

Figure 23.2: Main Smith Chart Axes

this thing apart piece by piece. We'll start with the big circle around the outside. You can see it highlighted in Figure 23.2a.

That's the *reactance axis*, and any point on it represents a particular reactance. Inductive reactances are on the upper half of that circle, capacitive reactances on the lower half.

Then we'll add an axis right across the center that represents resistance. Sensibly enough, it is known as the resistance axis. You can see it highlighted in Figure 23.2b. No capacitive reactance, no inductive reactance, just pure resistance. It runs from zero ohms on the left – a short circuit – to infinite ohms on the far right – an open circuit.

As you will see, the "circle with a line through the middle" part of a Smith Chart – with no other markings – can be used to solve some very useful problems.

Now, these two axes are all well and good except we seldom encounter transmission lines or antennas with zero or infinite resistance. There are resistance circles to represent constant real-world values of resistance. They all intersect that infinite resistance point on the right. (They don't look like they do because other circles get in the way, but they do.) Figure 23.3a shows a few of them highlighted.

There's another set of circles on the chart, but we only see part of the arcs of those circles, because the values represented by the parts of the circle outside the chart are impossible. You can see those arcs in Figure 23.3b.

Those are the *reactance arcs*. The reactance arcs in the top half of the

148

(a) Resistance Circles   (b) Reactance Arcs

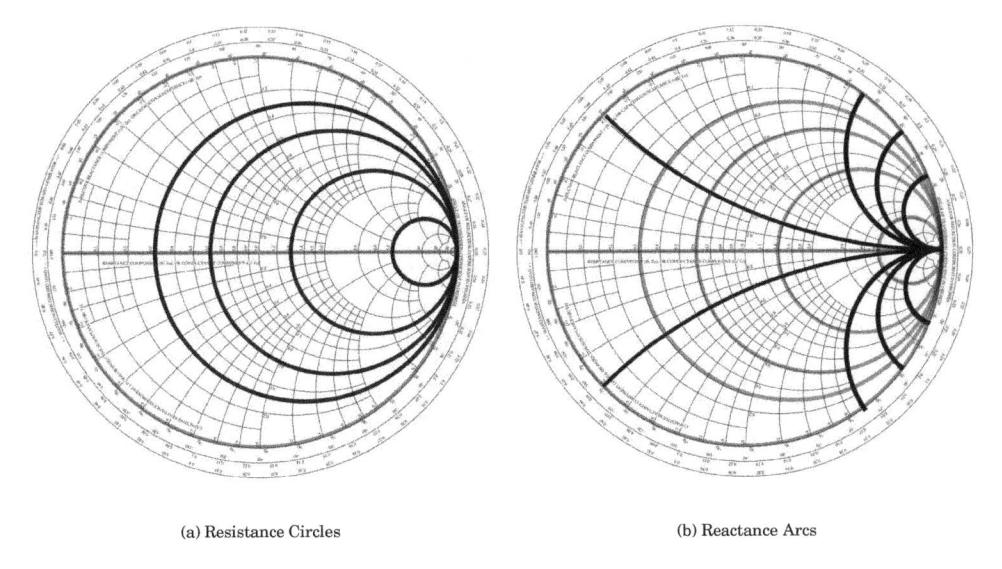

Figure 23.3: Resistance Circles & Reactance Arcs

chart – above the resistance axis – represent inductive reactance, the arcs in the bottom half are capacitive reactance.

There are also two wavelength scales on the Smith Chart. They run around the outside of the reactance axis, and they are calibrated in fractions of transmission line electrical wavelength. There are two scales because one runs clockwise around the circle and is for measuring "wavelengths toward a generator" (for us, in most cases, that means a transmitter) and the other for "wavelengths toward a load."

Finally, there are some scales across the bottom of the Smith Chart. Those are used to assign values to the results of various calculations, typically by drawing a line straight down from some particular point on the circular part of the chart or by using calipers to transfer a distance from the upper chart to the scales.

The "shape" and the spatial relationships of the lines on the Smith Chart are universal – they'll apply no matter what the values. If we could computerize the paper the Smith Chart is printed on (which is precisely what a network analyzer does) then the basic values could be set to whatever we want – but to make it work on paper, we have to have a common starting point. In effect, we set the chart up with the assumption that one part of the system has a fixed value. That value is the nominal impedance of the transmission line in the system, and on this chart – and most Smith Charts – it's 50 ohms. That value goes at the "prime center" of the chart, dead center on the horizontal resistance axis, a spot that is marked 1.0. That process of setting that fixed value is

149

called "normalizing the chart."

Because a 1.0 on the chart represents, for our purposes, 50Ω we also have to normalize all the values we plug into it. Very simple – divide everything by 50. (Cable TV system engineers divide everything by 75, since their systems use 75Ω impedances.) We'll demonstrate in a bit.

Figure 23.4: Slide Rule

With values plugged in, the Smith Chart becomes something very much like an old-fashioned slide rule like the one in Figure 23.4.

Application

Let's take a look at a few examples of how the Smith Chart can be used. Just to be clear, this is not "here's every way to operate the Smith Chart machine", this is "here's a brief look at what this machine looks like in operation."

We'll start with a simple example that happens to also have some very practical applications.

Coaxial Cable Stubs

What type of impedance does a 1/8-wavelength transmission line present to an RF generator when the line is shorted at the far end?

An *RF generator*, for our purposes, is a transmitter.

A 1/8 wavelength transmission line that is shorted at the far end presents **an inductive impedance**. Here's how we know that. All that is needed is a very, very simple version of the Smith Chart, like the one in Figure 23.5. You don't need any of the arcs, nor any of the numbers; just the left and right ends of the resistance axis, and the big reactance circle around the outside, plus the knowledge that one-half wavelength equals one full trip around the circle, so $\frac{1}{8}$ of a wavelength equals $\frac{1}{4}$ of the way around that circle going, for all these sorts of problems, in a clockwise direction.

Why would one trip around the circle equal one-*half* wavelength? So far as a transmission line knows, a positive half-cycle is no different from a negative half-cycle; both will encounter the same impedance.[2]

---

[2]While it is of little practical use to most of us, the exact characteristic impedance of a transmission line only appears at each half-wavelength interval. If you had the option,

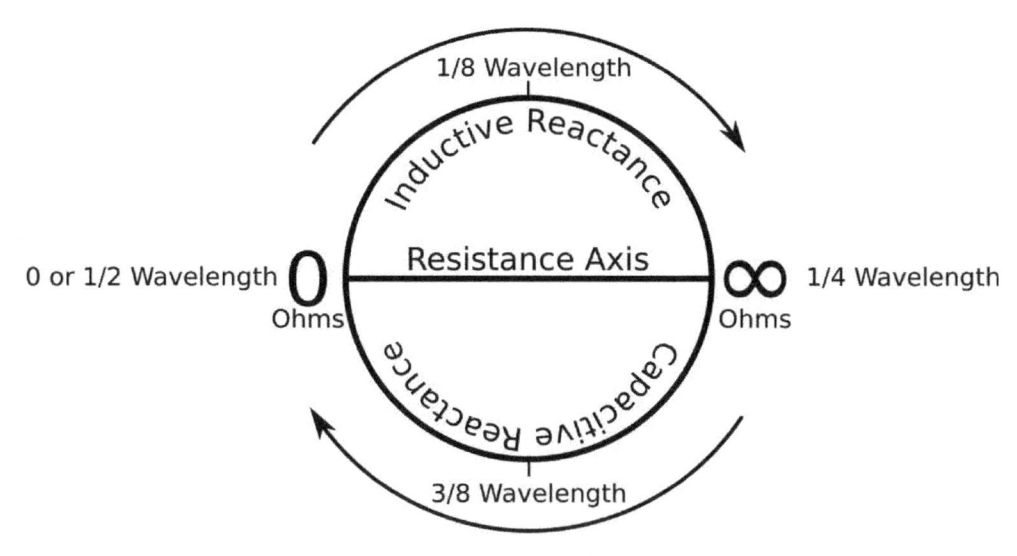

Figure 23.5: Super-Simple Smith Chart

It's also important to know that all impedances in the top half of the circle have inductive reactances and all impedances in the bottom half of the circle have capacitive reactances. We don't even need to normalize any values; we'll start either at the zero end or the infinity end of the resistance axis, depending on the given resistance value, then travel around the reactance axis a distance equal to the length of the coax stub in wavelengths.

We know the resistance of the piece of coax is zero ohms – it's shorted at the end. If we put our ohm meter on it, it will show zero ohms. (Or close enough for practical purposes.)

That puts our impedance somewhere on the reactance axis – the big circle around the outside, since it intersects the resistance axis at zero and, since it will come into play shortly, at infinity as well.

Now we can figure out the relative reactance. Follow along on Figure 23.5.

The wavelength scale goes around the outside of the chart. On a real Smith Chart there are some small notations with arrows that show us that for "wavelengths toward generator" we go around the scale clockwise. We start at the far left end of the resistance scale – zero ohms – and follow around to 0.125 (1/8) wavelength. 1/8 wavelength is ¼ of the way clockwise around the chart.

you'd cut your transmission line precisely at the half-wavelength point. Since most of us insist on having the capability of operating on more than one band (wavelength) and on multiple frequencies within those bands, this is a moot point.

Notice, that impedance point is in the upper half of the circle – and that means it has an **inductive**, +j reactance.

Let's try another.

What type of impedance does a 1/8-wavelength transmission line present to an RF generator when the line is open at the far end?

A 1/8 wavelength long transmission line with an open end presents a **capacitive impedance**.

This time we have an infinite (open circuit) resistance number, so rather than starting at the far left of the resistance line, we start on the far right; the infinity end of the resistance line. We trace clockwise around that reactance axis until we reach a spot opposite 0.125 wavelength, ¼ way around the circle, going clockwise.

We end up in the lower half of the circle, so it is an impedance with **capacitive reactance**.

What happens if we increase the length to $\frac{1}{4}$ wavelength?

What type impedance does a 1/4-wavelength transmission line present to an RF generator when the line is shorted at the far end?

Rather than making a quarter of a lap around the circle, we'll make a half a lap.

Since this line is shorted at the end, we'll start at the zero ohms mark, go halfway around the circle, and see that the answer is, according to the chart, infinite impedance. In reality, of course, we'd just call it a very high impedance. It's also a purely resistive impedance because we end up right at the "equator" of the chart.

What if we step up the length to $\frac{1}{2}$ wavelength?

What impedance does a 1/2-wavelength transmission line present to an RF generator when the line is shorted at the far end?

In the case of a 1/2 wavelength transmission line that is shorted at one end, we start at the zero ohm mark on the far left, go all the way around the circle, and end up at an impedance that looks just like the 1/4 wave open-end coax – mighty close to zero, with no reactance.

SWR Circle

We want to attach a load – let's say it's an antenna – to our 50 ohm coax. We know the resistance and the reactance of the load – those same values we used to create our phasor diagram.

(a) Plotting 50 Ω on the Resistance Axis

(b) Tracing the Resistance Circle

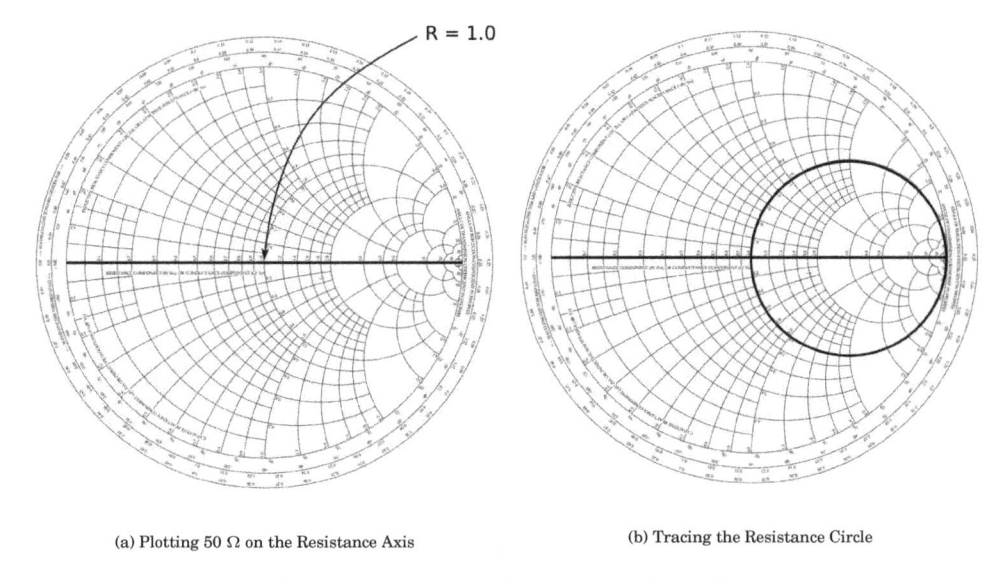

Figure 23.7: Plotting the Resistance Part of Z

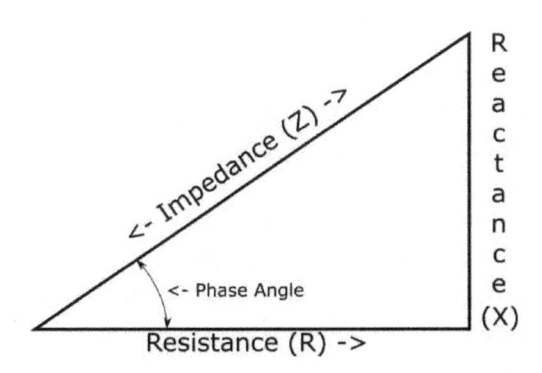

Figure 23.6: Phasor Diagram

What we want to know is what the SWR will be with that setup.

To keep numbers simple, let's say our load has a 50 ohm resistance and 50 ohms of inductive reactance. Formally, those are coordinates of (50, +j50.)

Step 1 is to normalize our values. We divide the resistance and the reactance by the normalizing number of the chart, which is 50. That gives us (1, +j1.)

Next, we plot the resistance point on the resistance axis. It's easy enough to find, 1.0 is smack dab in the center of the resistance axis, as shown in Figure 23.7a.

After that, we grab our gigantic #2 pencil and trace the resistance circle that crosses the resistance axis at that 1.0 point. You can see the

153

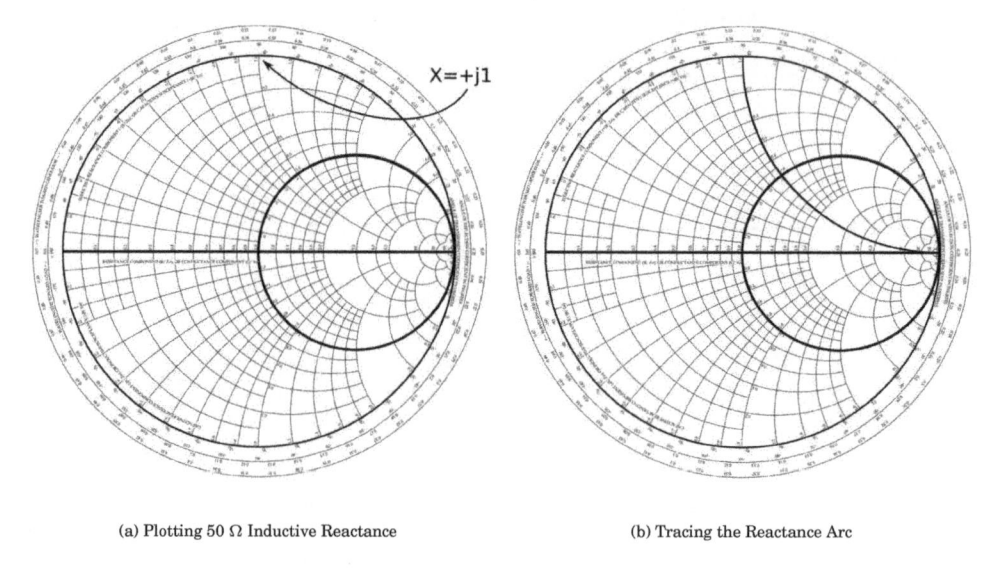

(a) Plotting 50 Ω Inductive Reactance (b) Tracing the Reactance Arc

Figure 23.8: Steps 2 & 3 Creating SWR Circle

result in Figure 23.7b.

That circle shows a resistance of 1.0 (normalized) ohms at every point. It doesn't really mean much of anything yet, though – we still need to figure the reactance point.

Next, we add the reactance point on the reactance axis. The reactance axis is that circle that marks the outside border of the chart. You probably can't see the numbers, but there are (tiny) numbers showing the values just inside that outer border. Negative numbers are on the bottom half of the chart – those are the capacitive reactances, the $-j$ numbers. Our reactance is 50 Ω of +j, so we'll look on the top half of the chart....and there it is, in Figure 23.8a. 1.0.

Next, we trace that reactance arc with our gigantic #2 pencil, as shown in Figure 23.8b.

Every point on that arc has a reactance of +j1.0. That spot where the resistance arc and the reactance arc meet is the (normalized) impedance of our load, which we can read on the chart as .... as .....well, the chart doesn't tell us the impedance in ohms. But we know that's a simple calculation, anyway, and that wasn't our question. We wanted to know something practical; what will the SWR be?

That point where the arcs meet is just like the point on our phasor diagram triangle where the hypotenuse meets the opposite side. See Figure 23.9.

We grab our handy compass, the kind you had in geometry class, not

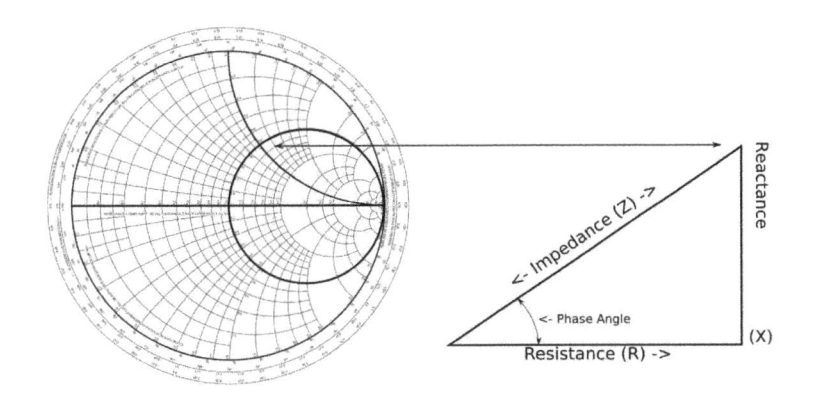

Figure 23.9: The Impedance is Where The Circle and the Arc Meet

the kind you'd use to find your way home from your campsite.

We put the point of the compass right in the center of the chart, at 1.0 on the resistance scale. Then we set the pencil on the impedance point, and draw a circle, as seen in Figure 23.10. What we have added is called a standing wave ratio circle. We read the SWR on the resistance scale. (We read it on the right hand side, to the right of 1.0, because there's no such thing as a SWR of less than 1:1.) I read it as 2.6:1.

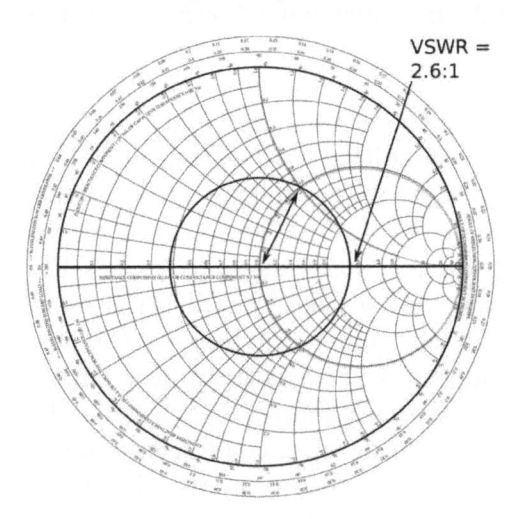

Figure 23.10: Drawing the SWR Circle

There are *many* more things you can calculate using a Smith Chart. If you'd like to explore a bit more, check out the *Fast Track* video on the topic at the AF7KB YouTube channel (http://tinyurl.com/AF7KB.) A professor at Humber College named Carl Oliver made a great set of videos on the topic on YouTube. Get started with this one: https://tinyurl.com/ppx89ng.

There's also a whole site devoted to the Smith Chart created by the Smith Chart Amateur Radio Society:

http://smithchart.org/phsmith.shtml

Introduction to VNA's

There was a time when owning a Vector Network Analyzer[3] was far beyond the realm of possibility for most hams. They are made by companies like Tektronix or Fluke; companies known for making professional equipment of superb accuracy and durability with stratospheric prices to match. For the applications in which those machines were – and still are – used, they are a bargain. They're also massively over-engineered for our ham shacks.

Today, the NanoVNA is available. It has limited range, limited functions, and a limited number of "ports." Comparing its performance and accuracy to a professional grade instrument is like comparing a plastic tricycle's performance to a Ferrari's but it happens to perform functions that are useful for us at an acceptable level of accuracy and at a very low price.

Put simply, VNA's measure what happens when you put a signal into a network. For us, that "network" is usually our transmission line and antenna. The VNA will tell us the resonant frequencies of that system and the nature and value of the impedance of that system. Using that information, it can also tell us the SWR at various frequencies. It does all this almost instantaneously, sweeping through a range of frequencies and plotting the resulting information in various ways.

One common analogy for what a VNA does likens a network to a train tunnel. If you posted a friend at the other end of the tunnel with a very accurate stopwatch, you could measure the length of the tunnel by shouting into the tunnel and measuring how long it took for the sound to reach your friend. (Working out the logistics of this plan is left as an exercise for the reader.) The science of vector network analysis uses *S parameters*" to define where the signal is being put into the system and where it is being measured. In the language of VNA's, that test where we holler down the tunnel to our friend on the other end would be an "S12" measurement. Not "ess twelve"; "ess one two." the "1" is the input "port" and the "2" is the output port.

If we yell down the tunnel and time how long it takes for the echo (if any) to get back to us, that's an S11 measurement. The input and output

---

[3]Sometimes known as a Network Vector Analyzer or even a Cable Analyzer.

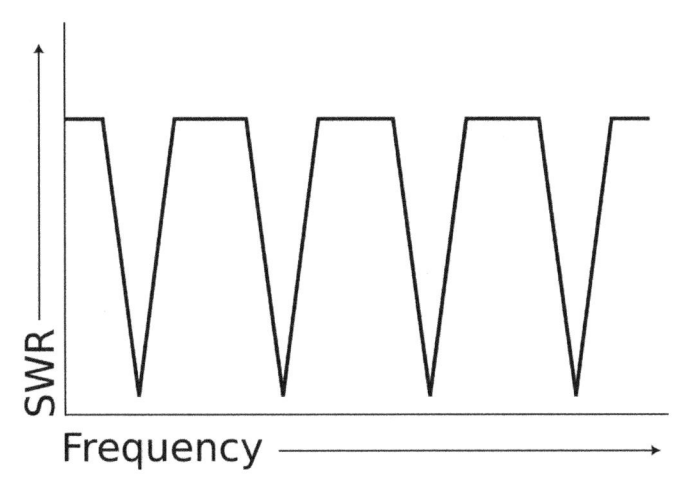

Figure 23.11: SWR vs. Frequency Plotted by VNA

ports are the same. That's what we're measuring when we measure SWR; when we put a signal into this system, how much of that signal reflects back to the input? In fact, all the common measurements we, as hams, might make with a VNA are S11 measurements.

One way to use a VNA is to find all the resonant frequencies of an antenna. In this case we're using the VNA as smart antenna analyzer. Rather than checking one frequency at a time, the VNA sweeps through a range of frequencies and displays the results. The graph you get will resemble Figure 23.11.

In Figure 23.11 you see the SWR taking pronounced dips at various frequencies as the frequency being tested passes through the antenna's resonant frequency. This is typical; antennas usually have multiple resonant frequencies.

We can also set the VNA to plot that same information on a Smith Chart. In that case we might get a picture that looks something like Figure 23.12.[4]

One useful way to read that display is by starting at the center of that spiral. First, recall that each point on that spiral marks an impedance at a different frequency. The location of the point relative to the resistance axis running across the center of the circle tells us if the impedance is capacitive, inductive, or purely resistive. The parts of the spiral above the equator are inductive, those below are capacitive. The point at the center of that spiral is the point at which the impedance is purely resistive with a value of 50 Ω. In other words, that's what you might call the sweet spot of this antenna.

[4]We really do mean "something like" – that's an idealized display.

157

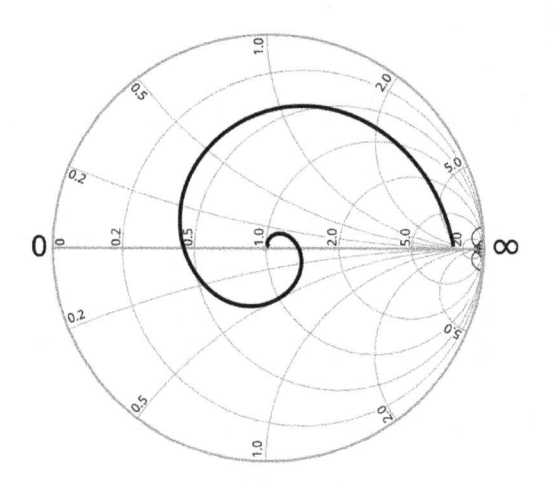

Figure 23.12: VNA Smith Chart Plot of Multiple Resonances

The other spots where the spiral crosses the resistance axis are the other resonant points of the antenna, though the impedance at those points is not 50 Ω. Unlike the paper Smith Chart, a VNA will let you identify exactly what frequency was being swept at each point on that spiral – or whatever shape your particular setup produces. On the NanoVNA it is simply a matter of poking the appropriate spot on the touchscreen display.

Chapter 24

Maxwell's Equations

We do not offer this chapter to you with the intention of having you learn to apply Maxwell's Equations to your ham shack. There will be no Application section of this chapter. Instead, we offer it for your admiration, much as you might admire the craftsmanship and clever design of a fine Swiss watch. After all, you do not need to master applying perlage, anglage, geneva stripes, and other decorative touches to a watch movement in order to appreciate that watch. However, if one knows what those things are and knows at least a little about what is required to produce them, one's appreciation and enjoyment increases.

You will be hard-pressed to find many pieces of mathematics with a higher "symbols-to-impact ratio" than Maxwell's Equations. They are, quite literally, the bedrock foundation of every telecommunications marvel (or plague, depending upon your point of view) of our modern world, and that is only part of their consequences. They show up in many other areas of physics including optics and relativity. Yet, they are not even mentioned in any ham radio exam we have seen!

Figure 24.1: James Clerk Maxwell

James Clerk Maxwell (1831 – 1879) was a professor at University of Cambridge when he created his famous Equations, published as *A Dynamical Theory of the Electromagnetic Field* in 1865. It is no exaggeration that among the giants of physics he stands shoulder-to-shoulder with Isaac Newton and Albert Einstein. It is said that Einstein kept pictures of three scientists in his study; Newton, Faraday, and Maxwell.

We're not sure exactly how Maxwell got on the path to the Equations. His interests were wide-ranging, encompassing everything from how gases behave to the nature of the rings of Saturn and even to the structural integrity of bridge trusses. He was certainly influenced by his admiration for and friendship with Michael Faraday. No one can dispute that Faraday was brilliant, but he never learned higher mathematics. Most of his discoveries are described qualitatively. Some historians think Maxwell's journey to the Equations began with his attempts to turn Faraday's prose into mathematics. In any case, Faraday's Law of Induction became an integral part of the Equations.

Figure 24.2: Charles-Augustin de Coulomb

Faraday's work did not spring from a vacuum. He had been preceded by a few signficant researchers. They included Charles-Augustin de Coulomb (1736 – 1806), André-Marie Ampére (1775 – 1836), and "the prince of mathematics", Carl Friedrich Gauss (1777 – 1855). Each of these contributed a crucial piece of the puzzle, but none achieved what Maxwell achieved by putting all their work together into a unified theory of electromagnetism.

We should point out that when Maxwell published his Equations, there were 21 of them. It took Oliver Heaviside to condense them down to four, though, to be fair, he created a set of four *integral* equations and four *differential* equations. Each set says essentially the same things, but one form is useful for some calculations and the other for other calculations. Technically, then, this chapter is about the Maxwell-Heaviside Equations.[1]

Figure 24.3: André-Marie Ampére

We will be using the *differential* forms of the Equations. The most common other form is the *integral* form. The Equations in integral form look something like Equation 24.1.

---

[1] Oliver Heaviside's persistent absence from popular science literature may relate to his, to put it delicately, personal quirks. He seems to have started off as somewhat anti-social, and this tendency was exacerbated through his life by worsening deafness. His commitment to personal grudges was legendary, as well.

$$\oint_S \mathbf{B} \cdot d\mathbf{A} = \mu_0 \left(I_{\text{enc}} + \epsilon_0 \frac{d\Phi_E}{dt} \right) \tag{24.1}$$

For reasons that should be understandable, most college courses begin the study of Maxwell's Equations with the simpler – or, at least, simpler looking – differential form of the equations. We will follow suit.

Here is an overview of what those four equations have to say.

- Gauss's Law for Electricity, also known as Coulomb's Law, states that electric charges produce an electric field.

- Gauss's Law for Magnetism says there are no magnetic monopoles; instead, magnetic field lines form closed loops.

  (One cannot help but note in passing that for any scientist, having one scientific law named for them would constitute a pinnacle of success. Gauss has two laws named after him. His IQ is estimated to have been at least 250 and perhaps as high as 300.)

- Faraday's Law of Induction describes how a changing magnetic field can induce an electric field.

- Ampére's Law with Maxwell's Addition shows how magnetic fields are generated by electric currents and changing electric fields.

You could be forgiven for feeling underwhelmed by these statements. Especially for those of us who have studied electromagnetism even casually, these equations declare things that are almost obvious to us; but, that's the point. They are inescapably true and they are *simultaneously* true. They allow no exceptions, no special conditions; so far as we know, except possibly in very, very strange conditions that do not occur on this planet, these rules are as close to absolutes as you will find in science.

It is probably not intuitively obvious to you how these fairly simple rules lead to, among other things, radio waves and an explanation of the nature of light. It was not intuitively obvious to Maxwell, either.

Let's look in a bit more detail.

Gauss's Law for Electricity

$$\vec{\nabla} \cdot \vec{E} = \frac{\rho}{\epsilon_0} \tag{24.2}$$

This is the law that says that electric charges produce an electric field. It doesn't say that directly, though.

First some basics about vectors. That arrow over $\vec{E}$ tells us we are dealing with a vector.[2] We touched on vectors when we talked about polar coordinates back in Chapter 17. There we mentioned that a vector has a direction and a magnitude. Put more simply, it is an arrow with a specific length pointing in some direction. It also has a location, which can be specified by x, y, and, if needed, z coordinates. A *vector field* is just a bunch of related vectors.

Figure 24.5 shows a vector field of the winds in some imaginary land. Obviously, the arrows indicate the direction the wind is blowing; toward the bottom of the map they are coming in from the west, then turning north as they encounter the mountains to the east. The length of the arrows indicates the speed of the wind. In a vector field, *every* infinitesimally tiny point has an associated vector. On paper, of course, we can only show a few of those, but the equations can deal with all of them.

Figure 24.4: Carl Friedrich Gauss

Any of the usual operations can be performed on vectors; addition, subtraction, etc. There are also special operators that apply only to vectors.

Mathematics uses vector fields to represent all sorts of real-world things. More importantly for our discussion, though, sometimes vectors are just vectors. You see, Maxwell rather brilliantly sidestepped the pesky question of "what, exactly, *is* an electric (or magnetic) field?" The equations offer us no information about that at all! The equations just talk about vectors and *fluxes* that behave in certain ways defined by the equations. ("Flux" is a measure of how much of a field is passing through a surface.)

In the map of the winds, we're representing the movement of air. "What's moving? Easy, air!" On another level, we're representing the *forces* that are causing the air to move the way it does. Those forces are revealed by the air's pattern of movement.

---

[2] An alternate way of showing that a value is a vector is by showing it in **boldface** and you will see that in many texts because it is difficult to typeset $\vec{E}$.

162

The vectors described by the Maxwell-Heaviside equations are similar. They tell us what the forces present must be by describing their effects on certain objects in those fields. In the case of the electric field, the object is a positively charged particle. In the case of the magnetic field, the object is another magnet such as a compass needle. Those objects respond to the fields, and their response is what the vectors describe. Consider this; when we reveal a bar magnet's magnetic lines of force, the iron filings don't keep moving around on the paper even though the vectors that describe that force point out of the north pole of the magnet and disappear into the south pole. The vector lines represent a force, not motion.

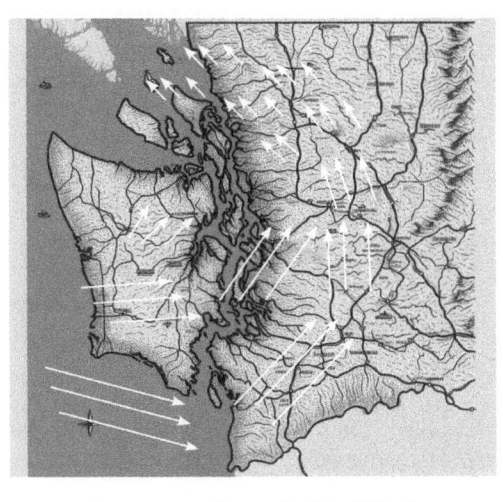

Figure 24.5: Vector Field of Winds

Gauss did a little sidestepping of his own with his Law for Electricity. Rather than attempt to define the nature and precise location of an electric charge which is, after all, a target that is invisible, moving, and expanding, he said, with this equation, "Build a sphere around an electric charge. *It doesn't matter how big the sphere is*. The surface of that sphere will always contain precisely the same amount of electric charge as the charge it encloses."

As you might expect, mathematics has developed many ways to mathematically describe various behaviors of vectors. Aside from an individual vector's magnitude and direction, we can also describe the behavior of the entire vector field or of regions of that field.

Since we are dealing with rates of change in complex systems with infinite numbers of elements, we have now arrived solidly in the territory of calculus. We will do our best to explain the concepts that are at play in these equations in simple language, but the truth is we will hit a point in this chapter where the maths has gone beyond simple explanation and into the world of upper-level calculus. We will warn you in advance: At that point, we're going to wave our hands and mumble something like, "And then a miracle happens."

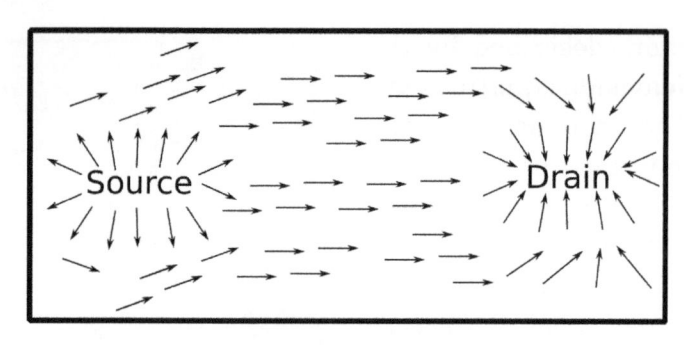

Figure 24.6: Vector Field

The $\vec{\nabla} \cdot \vec{E}$ index$\vec{\nabla}\cdot$ part of the equation says, "the divergence of the vector field called E (the electric field) equals ρ, the charge density, divided by ϵ_0, the permittivity of free space, also known as the electric constant.

There is no reason to memorize the permittivity of free space, but there are two important things about it. One is simply that the right-hand side of the equation, $\frac{\rho}{\epsilon_0}$, is not zero. It could be negative or positive. (If it is zero there is no field so the equation becomes meaningless; 0 = 0.)

The other important thing is a pair of values in the equation of that permittivity number; the m and the s.

$$\epsilon_0 = 8.85418782 \times 10^{-12} \times m^{-3} \times s^4 \times A^2 \qquad (24.3)$$

The m in that equation stands for meters and the s stands for seconds of time. Those will turn out to be key elements in our story. (For the curious, the A is for area.)

Divergence is a vector mathematics term that is concerned with the destinations of those vector arrows within a specified region.

Imagine a shallow tank with a water source – like a bathtub spout – at one end and a drain at the other. We'll also imagine a slight slope from the source end to the drain end.

If we made a vector field of the flow of water in our tank, it should look something like Figure 24.6.

We'll make this an ideal tank, where every atom of water that comes out of the faucet eventually goes down the drain. The *divergence* of the vectors in the entire tank, then, is zero.

Mathematicians would say the divergence of the vectors in the left $\frac{1}{3}$ of the tank, right around the source, is positive. The vectors are flowing out of that point and they never (in our ideal tank) come back.

The vectors in the middle of the tank are positive, too.

The vectors around the drain are negative; they are all flowing toward the drain.

As you can see, we cannot talk about the divergence of a vector field without identifying whether we are talking about a part of the field or the entire field. In all of Maxwell's Equations, we are considering the entire field.

We said that $\frac{\rho}{\epsilon_0}$ is not necessarily zero, so Gauss is telling us that the force of the electric field would push a positively charged particle outward, and since we can make the sphere that intercepts that field any size we want and still end up with the same amount of charge on the surface of the sphere, then that must mean *that field will radiate outward forever.* If the charge creating the field is negative,[3] the forces point inward, toward the charge, but the field still radiates outward. No matter what part of the electric vector field we look at, the divergence is either positive or negative.

In the everyday world, this infinite expansion is quite unusual. If you blow up a balloon, then release the air, the balloon deflates – pressure went in, then pressure went back out because the balloon resisted the pressure. Gauss tells us that for electric fields *there's no balloon.* In free space (the ϵ_0 part of the equation) nothing reduces that charge and nothing stops its outward expansion. It radiates out from some point inside that sphere and never comes back.

Gauss's Law for Magnetism

$$\vec{\nabla} \cdot \vec{B} = 0 \qquad (24.4)$$

Now that you have some insight into vector maths, you can probably see this equation bears some resemblance to Gauss's Law for Electricity, $\vec{\nabla} \cdot \vec{E} = \frac{\rho}{\epsilon_0}$. Here we're talking about $\vec{B}$, the magnetic vector field, rather than $\vec{E}$, the electric field and Gauss tells us the divergence ($\vec{\nabla}\cdot$) of the magnetic field $\vec{B}$ is *zero*! Just like our water tank, everything goes down the drain!

Consider the shape of the magnetic field around a bar magnet. Most

---

[3]This would mean ρ, the charge, is a negative number. In vector calculations, making a vector negative reverses its direction.

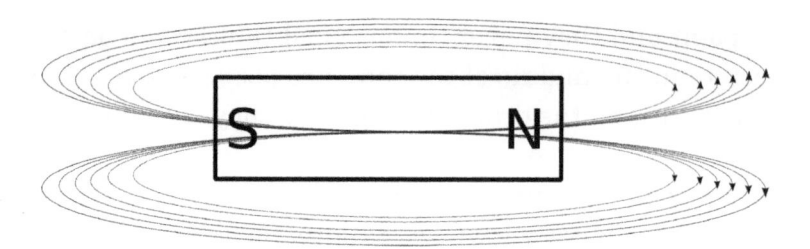

Figure 24.7: Vector Magnetic Field

of us have seen the experiment where we place a bar magnet under a sheet of paper and sprinkle on some iron filings to reveal the magnetic field. It shows the shape is something like you see in Figure 24.7.

In that vector diagram, the north pole of the magnet is the source and the south pole is the drain. In the language of mathematics, it is the *sink*. Gauss's Law for Magnetism tells us that the divergence of those vectors is zero; in other words, every single magnetic line of force that comes out of the north pole goes back into the magnet at the south pole. Therefore, there can be no such thing as a "magnetic charge." In fact, there's no such thing as a "starting point" for a magnetic line of force – it just loops around and around.

Something like a neutral atom, a magnet is *overall* neutral. An electron, such as those we find floating around on their own in the ionosphere, is electrically negative. It is a "monopole." Obviously, electric monopoles can exist. No "magnetic monopoles" can exist.[4]

For us humble practitioners of the dark arts of electromagnetic radiation, this tells us that when we create a magnetic field around an antenna – and we'll shortly see an equation that tells us we do that every time we pass a current through it – the magnetic field radiates out one end then back into the opposite end of the antenna! Picture, then, the electric field radiating straight out from the antenna like the spokes of a bicycle wheel with the antenna as the axle while the magnetic field radiates along the axis of the axle. This explains why we say the electric field and the magnetic field are polarized 90° from each other. In fancy physics terms, they are *orthogonal*.

Some of the profound things the Maxwell-Heaviside equations say

---

[4]Of course, the discovery or production of a magnetic monopole would be Nobel Prize contender physics, and there are people working on creating such a thing. Here in the real world, Gauss is still absolutely right.

are really things they *don't* say. Gauss tells us the magnetic field has zero divergence *everywhere*; there are no location coordinates for the $\vec{B}$ field. There is no size limit in that equation. Like the electric field, the magnetic field goes out forever. Of course, just like the electric field, the strength of the field at any one point diminishes with distance as the field spreads out across more space.

Something else profound that the equation does not say is that the magnetic field must be attached to an object. For that matter, neither does Gauss's Law for Electricity require the electric field be attached to an object.

Faraday's Law of Induction

$$\vec{\nabla} \times \vec{E} = -\frac{\partial \vec{B}}{\partial t} \tag{24.5}$$

Here is where the electric field and the magnetic field are linked.

We learn about induction in our earliest days as hams. In transformers, for instance, passing an alternating current through a coil that is adjacent to another coil induces a current in the second coil.

In this equation we meet a new vector mathematics term, $\vec{\nabla}\times$. That is the *curl* of the vector field. Curl is another term used to describe what regions of vectors are doing. Formally, it is a measure of how much the vectors "circulate" around a point with respect to time. (Please notice that time just stuck its nose under the tent, again.) *Circulate* has a very technical definition in the world of vector mathematics with which we are not going to concern ourselves; we can just think of it as "change." In plain language, Faraday's Law simply says the *rate of change*, $\frac{\partial B}{\partial t}$ of the magnetic field is equal to the curl (another rate of change) of the electric field. (∂ means "partial differential"; for our purposes, just read it as "rate of change.") Even more simply, changing the vector of the electric field creates a magnetic field that is proportional to the rate of change of that field.

Notice the negative sign on the right-hand side of the equation. that has to be there; otherwise when you switched on a transformer it would produce infinite energy and destroy the universe. People would get upset.

With this equation we see a new symbol; ∂. That indicates a *partial differential*. That means we have stepped bravely into the realm of calculus and, particularly, *differential calculus*. If you never took calculus, or took it but have either forgotten or never grasped its fundamental ideas, it turns out that those ideas are really rather simple. Calculus is daunting because of the methods of implementing those ideas; they are sort of the advanced version of long division's "divide the divisor into the dividend's first digits, multiply the answer by the divisor, subtract the result from the original digits, etc."

Differential calculus is concerned with the rate of change of one quantity with respect to another quantity. We're showing you the differential calculus forms of the equations in this chapter. Another form of Maxwell's Equations translates the Equations into *integral calculus*, which is like a mirror image of differential calculus; it is concerned with, essentially, how much of something accumulates at given rate(s) of change. Imagine a sloped line on a graph. Differential calculus can tell you the slope of the line at any given point, which is a pretty good trick since the point itself has no slope! Integral calculus can tell you the quantity of the area under the line, or a part of the line, and it can do it even if the line is irregular.

This is one place where Maxwell most likely intervened rather heavily in Faraday's original material. Faraday almost certainly did not express his law of induction in that formula. Faraday barely knew algebra. Calculus was far beyond his mathematical capabilities. So, this formula is Heaviside's translation of Maxwell's translation of Faraday's careful measurements and qualitative description of his experiments. Sometimes it takes a team to do heavy lifting.

What this Equation implies is that the electric field and the magnetic field are, somehow, interchangeable. What if it turned out that a change in the magnetic field created an electric field? (Spoiler alert; it does.)

Ampére's Law with Maxwell's Addition

$$\vec{\nabla} \times \vec{B} = \mu_o(\vec{J} + \frac{\partial \vec{E}}{\partial t}) \tag{24.6}$$

This is the final piece of the electromagnetic puzzle.

There are a lot of moving parts in this equation, so let's catalog them. The good news is we have seen most of these parts already.

- $\vec{\nabla} \times \vec{B}$ is the curl of the magnetic vector field, B. For Maxwell's Equations we can read "curl" as "change."

- μ_0 is the *permeability* of free space. Permeability is a measure of the effect a substance (or, in this case, lack of substance) has on a magnetic field that passes through it.

- J is the current density, representing the amount of current per unit area. This is not quite the same as ρ which is the *charge* density in an area, but it is certainly related.

- ϵ_0 is a term we have seen before. It is the permittivity of free space.

- $\partial \vec{E}$ is the electric equivalent of $\partial \vec{B}$ we just saw in Faraday's Law of Induction. Here it is the rate of change of the electric field.

- ∂t is the time element of this rate of change fraction.

Maxwell's Addition is the $\frac{\partial \vec{E}}{\partial t}$. He needed to add that to account for what is known as the *displacement current*. Displacement current is not the flow of charges in a conductor; that's called *conduction current*. Displacement current is, really, something like an accounting adjustment; it accounts for parts of the moving electric field where there is no actual current. Think of the current that seems to pass through a capacitor when it is charging or discharging. The dielectric in that capacitor is an insulator – there's no "real" current passing through it, yet if we measure the current on either side, we find real amperes. That is the result of displacement current as the charge on one plate displaces charges on the other plate.

Electromagnetic Waves

Perhaps you have noticed that thus far there has been not a single mention of electromagnetic waves in this chapter, even though we have covered all four of Maxwell's Equations. Why, then, are they renowned as the foundation of electromagnetism?

At the beginning of the chapter we mentioned that Maxwell's Equations are not only true; they are *simultaneously* true. The Equations tell us that a changing electric field creates a magnetic field. They also tell us that a changing magnetic field creates an electric field. They tell us that those fields expand out into space from the source. As they radiate through space, the electric field creates a magnetic field, which creates an electric field ... It is a self-propagating wave! Is that really what Maxwell's Equations tell us?

Getting to electromagnetic waves involves a bit more work with those equations. By "a bit more work," we mean that in order to explain the steps to you we'd need to first back up a big truck of advanced calculus books and dump the contents into your head. (To be honest, we'd probably need the first truckload for ourselves.) Maxwell published his final form of the Equations in his landmark paper in 1865, but he started working on them in the 1850's and continued working on them into the 1870's.

We won't go through the very high level vector calculus required to accomplish this. Seriously, it involves an operator that is "the divergence of the gradient of a function" among other exotic curiosities.

What the steps accomplish is to mathematically combine Ampére's Law with Faraday's Law.

At the end of a process filled with lots and lots of those $\vec{\nabla}$'s and ∂'s, what pops out is two equations that even undergraduate physics students instantly recognize as *wave equations*.

$$\vec{\nabla}^2 \vec{E} = -\mu_0 \epsilon_0 \frac{\partial^2 \vec{E}}{\partial t^2} \qquad (24.7)$$

$$\vec{\nabla}^2 \vec{B} = \mu_0 \epsilon_0 \frac{\partial^2 \vec{B}}{\partial t^2} \qquad (24.8)$$

The wave equations show that the electric and magnetic fields propagate as waves with the speed of light. Equation 24.7 shows the electric field wave, 24.8 the magnetic field wave. They are quite similar to the equations for sound waves or ocean waves.

Every equation can be represented graphically – every equation paints a picture. These wave equations paint pictures of three-dimensional waves traveling through space.

There's still nothing in those equations about meters-per-second, nor is there any other obvious element that would tell us speed. How did we get from those equations to speed-of-light?

The speed-of-light infiltrated our operation under cover of those constants, μ_0 and ϵ_0; the magnetic and electric constants.[5]

Maxwell reasoned that given the way the magnetic field and electric field interact with each other, there must be a mathematical relationship between ϵ_0 and μ_0.

With a bit more working on those equations, we find that we can actually derive the numerical value of the speed of light. It is what you see in Equation 24.11.

Recall that one of the equations includes ϵ_0, the permittivity of free space, and another includes the permeability of free space, μ_0. We said ϵ_0 is equal to $8.85418782 \times 10^{-12} \times m^{-3} \times s^4 \times A^2$. μ_0 is equal to $1.25663706 \times 10^{-6} \times m \times kg \times s^{-2} \times A^{-2}$. For this calculation, though, we'll use different, simplified forms.

$$\epsilon_0 = 4\pi \times 10^{-7} \tag{24.9}$$

$$\mu_0 = 8.854 \times 10^{-12} \tag{24.10}$$

By plugging in those values, we get 300,000,000 meters per second.

$$c = \frac{1}{\sqrt{\epsilon_0 \mu_0}} = \frac{1}{\sqrt{(4\pi \times 10^{-7}) \times (8.854 \times 10^{-12})}} = 2.997 \times 10^8 \frac{m}{s} \tag{24.11}$$

That means we can simplify the wave equation to an equation that does include c, the speed of light.

$$\vec{\nabla}^2 \vec{E} = \frac{1}{c^2} \frac{\partial^2 \vec{E}}{\partial t^2} \tag{24.12}$$

For Maxwell, it was not a long logical jump from that equation to realizing that these electromagnetic waves were, in fact, light. We can easily imagine that the moment he saw that relationship was a profound one for Maxwell. He was, after all, a man fascinated by light. He had done many experiments with spinning color wheels, studying how colors on the wheels combined under different circumstances. He had made the first durable color photograph in 1861. Here, he had revealed that the nature of light is that it is an electromagnetic wave; that it was the same "stuff" with which scientists from Volta to Faraday had been experimenting. He

---

[5] μ_0 and ϵ_0 are also known as the permeability and permittivity of free space, respectively.

showed that the force that created his color photograph was the same force that had made frog legs jump for Galvani and that charged Ben Franklin's Leyden jars.

James Clerk Maxwell did predict the existence of what we would come to call radio waves, but it would take some 22 years until Heinrich Hertz would prove he was correct and would show how to produce and detect them. We can probably blame the near-total lack of practical guidance contained in the Equations but can also point to the fact that no one had the slightest idea that such a thing would be useful. Hertz himself is said to have regarded his own experiments as of little practical value. One possibly true story claims Hertz was once asked what would be the use of his discovery. His response was, "I suppose nothing!"

His "nothing" turned out to be the key ingredient in a profound alteration of each of our lives. It is no wonder he could not imagine what has happened; who could have?

Index

τ, 81
$a^2 + b^2 = c^2$, 1
2π, 68
J, 169
Ω, 67
ϵ_0, 165
μ_0, 169
ω, 68
∂, 167
ρ, 165
τ, 81
$\vec{B}$, 169
$\vec{\nabla}\cdot$, 165
$\vec{\nabla}\times$, 167
ζ, 77
n, meaning, 4
xy graph, 4
$y = sin(x)$, 6
Ørsted, Hans Christian, xi
90-degree phase angle, 87

Al-Khwarizmi, Muhammad ibn Musa, 9
algebra, 9
algorithm, 9
Ampére's Law with Maxwell's Addition, 161
Ampére, Andre-Marie, xi, 160
Amplitude Modulation Bandwidth, 120
analog amplitude modulated signal bandwidth formula, 123
angular speed, 68
answer toggle key, 16
antenna analyzer, 157
azimuth, 100

balanced oscillator, 120
bandwidth, 117
bandwidth efficiency, FM, 123
bandwidth vs. noise, 115
bandwidth, AM signal, 123
bandwidth, FM, 123
bandwidth, ITU definition, 122
bandwidths, FM, 115
baseballs, 18
Bell Laboratories, 124
Buchanan, James, 136

calculator, recommended, xii
calculus, 1
capacitive reactance formula, 68, 96
capacitive reactance graph, 63
capacitor banks, power distribution, 92
Cartesian coordinates, 4
coaxial cable, 135
coaxial cable impedance formula, 139
coaxial cable structure, 137
coaxial cable stubs, 150
conductance, 42
convert standard values to metric values, 15
Coulomb, Charles-Augustin de, xi, 160
curl, 167
CW bandwidth, 118

damping ratio, 77
dB, usefulness of, 29
Death Star, 15
decibel power gains and losses, 30
decibels, 29

173

desmos.com, 8
deviation ratio, 113
dielectric, 138, 143
dielectric constant, 138
differential calculus, 168
Digital Signal Bandwidths, 124, 126
dimensionless unit, 29
distributed inductance, 143
divergence, 164
divide by zero, 11

Einstein, Albert, xvii, 159
EIRP, 129
elastance, 43
electric monopole, 166
electromotive force, 35
Electronic Applications of the Smith Chart: In Waveguide, Circuit, and Component Analysis (Electromagnetic Waves), 145
elevation, 100
ELI the ICE man, 90
Eliot, T S, 89
engineering notation, 14
engineering notation, TI-30XS, 16
equal sign, 1
ERP, 129
exponents, 14

Faraday's Law of Induction, 161
Faraday, Michael, 160
Faraday,Michael, xi
fluorinated ethylene propylene, 138
flux, 162
FM Bandwidth Calculations, 123
frequency from wavelength, 25

gain, 29
Gauss's Law for Electricity, 161
Gauss's Law for Magnetism, 161
Gauss, Carl Friedrich, 160
giga, 18
graphing, 4
gutta percha, 137

half-power bandwidth, 75, 76
Hartley, Ralph, 124

Heaviside, Oliver, xi, 136, 160
Hertz, Heinrich, 172
heterodyning, 51, 120

ideal component, 49
IF section, 52
image response, 51
impedance, 95
impedance formula, 97
impedance matching, 102
impedance of open 1/8 wavelength transmission line, 152
impedance of shorted 1/8 wavelength transmission line, 150
impedance of shorted ½ wavelength transmission line, 152
impedance, factors affecting, 95
inductive reactance formula, 67, 96
inductive reactance graph, 63
integral calculus, 168
intermediate frequency, 52
intermodulation distortion, 51
intermodulation interference, 57
inverting differential amplifier, 111
isotropic radiator, 129
ITU, 122

jellybeans, 18
Joule's Law, 37, 40

Kennelly-Heaviside layer, 89
keying, 118
kilo, 18
Kirchoff's Law of Voltage, 39

ladder line, 135
link budget, 131
link margin, 131
local oscillator, 52
logarithm, definition, 31
Lord Kelvin, 137
loss in a coaxial cable, 139

magnetic monopole, 166
Mathematical Operators, 2
Mathprint mode, 16
Maxwell's Equations, xvi, 159

Maxwell, James Clerk, xi, 159
Maxwell-Heaviside Equations, 160
mega, 18
metric prefixes, 13
micro, 18
Miller, Steve, 47
milli, 18
milliamps, 14
milligrams, 14
millivolt, 14
mode, TI-30XS, 16
modulation index, 113, 123
modulation index formula, 114

nano, 18
NanoVNA, 156
nanoVNA, 145
Narrowband FM, 124
necessary bandwidth, 117
Network Vector Analyzers, 100
Newton, Sir Isaac, 159
noise floor, 59
non-linearity, 57
normalizing the chart, 150
NTIA, 117
number line, 4
numbers, types of, 3
Nyquist, Harry, 124

Ohm's Law, xi, xiv, xvi, 35
Ohm's Law, reactance, 61
Ohm's Pie, 36
Ohm, George Simon, xi
Old Possum's Book of Practical Cats, 89
Oliver Heaviside, 89
online graphing calculator, 8
op-amps, 111
operators, 1
oscilloscope, 113

parallel circuit, 41
parallel resonant circuit, 62
peak voltage, 48
peak-to-peak voltage, 48
PEP, 129
per, 22
permeability, 169

phantom signal, 51
phase angle, 87
phase shifts, 61
phasor diagrams, 98
pico, 18
plenum, 138
polyethylene, 138
power factor, 30, 92, 130
Power Pie, 37
prime center, 149
proportional relationship, 10

Q factor, 73
Q formula, parallel, 74
Q formula, series, 73
Queen Victoria, 136

radians, 68
radical, 62
reactance, 61, 67
reactance arcs, 149
reactance axis, 148
reciprocal, 68
rectangular coordinates, 4, 62, 99
relation symbols, 1
relative permeability, 142
relative permittivity, 140, 142
resistance axis, 148
resistance circles, 148
resonance, 61
resonant circuit, 61
resonant frequencies, antenna, 157
resonant frequency formula, 62
ridiculous resistance problem, 42
ringing, 77
RMS voltage, 47
Root Mean Squared, 47

S parameters, 156
scientific notation, 14
selectivity, 51
sensitivity, 51
series circuit, 41
series resonant circuit, 62
series/parallel circuit, 41
Shannon, Claude, 124
Shannon-Hartley Theorem, 124

175

Shannon-Nyquist Theorem, 124
SI values, 13
sideband, 121
sine wave, 6, 87
slide rule, 150
Smith Chart, 146
Smith Chart web site, 156
Smith Chart, full, 147
Smith Chart, on VNA, 157
Smith, Phillip, 145
speed of light, 21
square root bracket, 62
square root of 2, 47
standard value, 18
standing wave ratio circle, 155
Star Wars, 15
starting capacitors, 92
stereo FM, 115
subscripts, 4
substations, electrical, 92
superheterodyne, 52
superscripts, 3
Système International, 13

Telegrapher's Equations, 89
The Electrician, 89

TI-30XS, xii
time constant, 81
time constant formula, capacitor, 81
time delay circuit, 82
total reactance formula, 97
trans-Atlantic telegraphy, 89
transmission lines for impedance matching, 141
trigonometry, 7

unmodulated carrier, 113

vector, 100, 162
vector field, 162
Vector Network Analyzer, 156
velocity factor, 137, 138, 142
VFO, 52
volt, 35
voltage divider, 39

wavelength, 21, 24
wavelength formula, 25
wavelength scales, 149
Wavelength Wheel, 25
Whitehouse, Dr. Edward, 136
window line, 135

Ohm's Law Wheel:

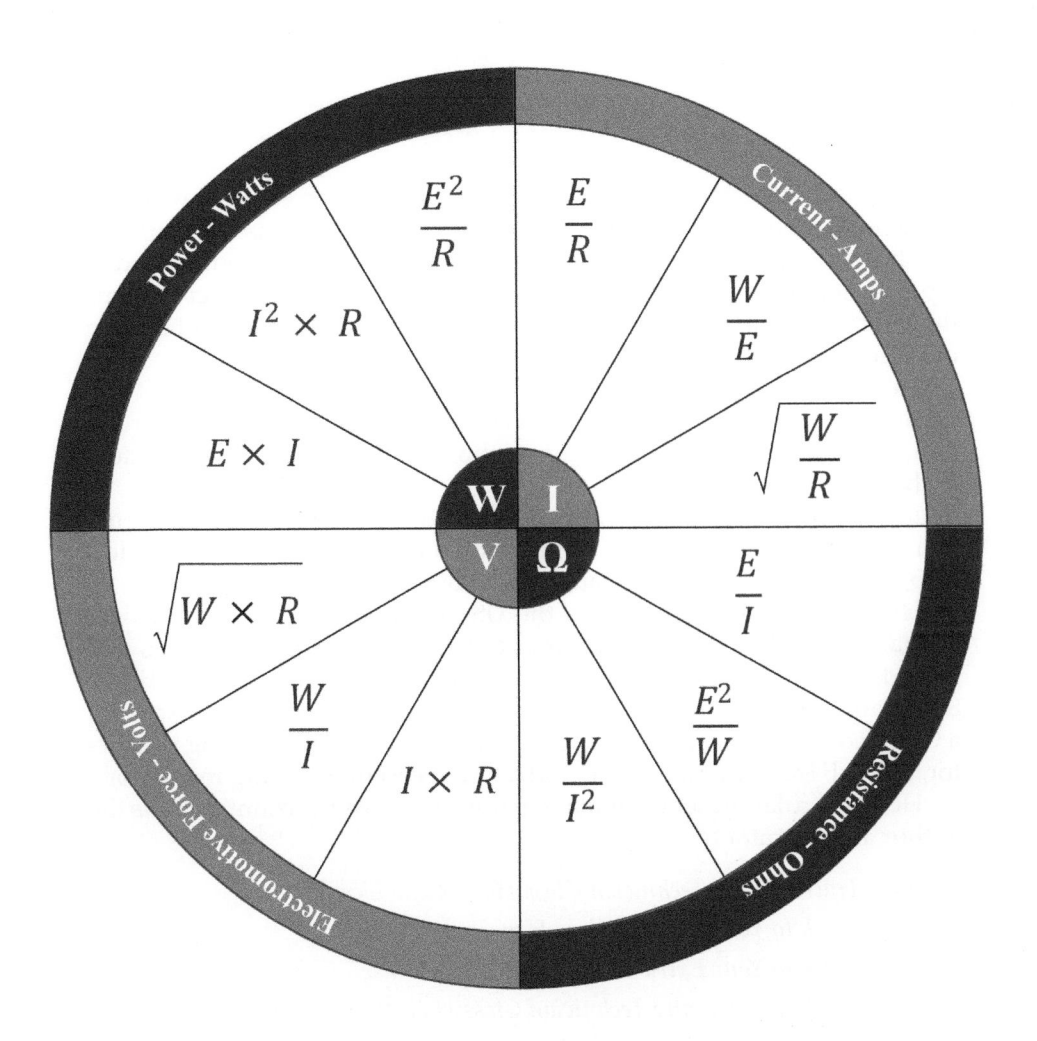

- Power - Watts: E^2/R, $I^2 \times R$, $E \times I$
- Current - Amps: E/R, W/E, $\sqrt{W/R}$
- Resistance - Ohms: E/I, E^2/W, W/I^2
- Electromotive Force - Volts: $\sqrt{W \times R}$, W/I, $I \times R$

Center: W, I, V, Ω

Do not worry about your difficulties in Mathematics. I can assure you mine are still greater.

- Albert Einstein

About the Authors

Michael Burnette, AF7KB, started playing with radios at age 8 when he found the plans for a crystal radio set in a comic book and wasted a half- roll of toilet paper to get the cardboard tube for a coil form. That radio failed as a practical appliance when it proved to only receive high-power stations that were less than one city block away. What would become a life-long quest for more effective communication had begun.

He spent some 25 years as a commercial broadcaster.

By 1989 he owned his own stations in Bend, Oregon, which afforded him abundant opportunities to repair those stations, often in the middle of the night in a snowstorm.

In 1992, Burnette left the radio business behind, and took to traveling the world designing and delivering experiential learning seminars on leadership, management, communications, and building relationships.

He has trained people across the US and in Indonesia, Hong Kong, China, Taiwan, Mexico, Finland, Greece, Austria, Spain, Italy, and Rus- sia. In addition to his public and corporate trainings, he has been a National S--ki Patroller and volunteer EMT, a Certified Professional Ski Instructor, a Certified In-Line Skating Instructor, a big-rig driver and driving instructor, an NLP Master Practitioner, and a Certified Firewalk- ing Instructor

He is a regular presenter at major hamfests across the country. He is the author and narrator of:

The Fast Track to Your Technician Class Ham Radio License

The Fast Track to Your General Class Ham Radio License

The Fast Track to Your Extra Class Ham Radio License

The Fast Track to Mastering Technician Class Ham Radio Math

The Fast Track to Mastering General Class Ham Radio Math

The Fast Track to Mastering Extra Class Ham Radio Math

The Fast Track to Understanding Ham Radio Propagation

The Fast Track Book of Ham Radio Facts

These days he makes his home in the Seattle, WA area with his wife and co-author, Kerry, KC7YL. Kerry was a high school maths and science teacher for over a decade and holds a Master of Education in Teaching as well as her Extra Class amateur radio license.

Find teaching videos, chapter-by-chapter practice exams, and more on fasttrackham.com.